The Random Walks
of
George Pólya

The Random Walks
of
George Pólya

Gerald L. Alexanderson

with essays by
Ralph P. Boas, Jr., Kai Lai Chung, D. H. Lehmer,
R. C. Read, Doris Schattschneider, M. M. Schiffer,
and Alan Schoenfeld

Published and Distributed by
The Mathematical Association of America

©2000 by
The Mathematical Association of America (Incorporated)

Library of Congress Catalog Card Number 99-67969
ISBN 0-88385-528-3

Printed in the United States of America

Current Printing (last digit):
10 9 8 7 6 5 4 3 2 1

SPECTRUM SERIES

MAA Service Center
P. O. Box 91112
Washington, DC 20090-1112
800-331-1622   FAX 301-206-9789

Preface

This biography would not exist had it not been for the urging of Peter L. Duren and Richard A. Askey, both longtime admirers of George Pólya and his work.

Next I am pleased to acknowledge the use of vast amounts of materials from the Pólya papers, now housed in the Stanford University Archives, and the gracious cooperation of Roxanne Nilan and Margaret Kimball, University archivists. I am grateful to Ellen Heffelfinger who searched through the Chowla Archives at the American Institute of Mathematics for correspondence concerning the Pólya conjecture. Sanford Segal of the University of Rochester was extremely helpful in leading me to some critical letters by Pólya, Weyl, and Hecke, currently in the hands of Niels Jacob of the University of Erlangen, who generously allowed them to be quoted here. Frank A. Smithies, knowing Cambridge as he does, was of great help in identifying some of the more obscure entries in G. H. Hardy's lists. Peter Lax read over the almost complete manuscript and caught a few slips in my descriptions of Hungary and Hungarians (those Hungarian accents can be very tricky!). He caught some errors and also contributed some anecdotal material of which I was unaware.

For similar contributions I am grateful to Lester H. Lange and Jean J. Pedersen, both of whom were close to Pólya and recalled incidents I never knew or had forgotten. For translations of the Hungarian materials I depended on my colleagues, Paul R. Halmos and Istvan Mocsy, the latter also for his large library of Hungarian reference materials. Istvan's patience in tracking down dates and geographical details about Budapest was boundless. For a careful reading and many suggestions, I am also grateful to Leon Bowden, who worked closely with Pólya on a couple of his books, to R. Lee Van de Wetering, to Donald J. Albers, and to Ina Lindemann. For translations of hard-to-decipher German script I relied on my colleague, Monika Caradonna. For the clever title, which I would never have come up with myself, I give credit to my friend, Constance Reid. And last but by no means least, I owe a great deal to Pólya's nieces, Susi Lányi and Judi Kennedy, who read the manuscript carefully and caught a number of errors, including, again, some of those pesky Hungarian accents. For the Mathematical Association of America, the final readers were Arthur Benjamin, Jennifer J. Quinn, and, again, Sanford Segal, who made many fine suggestions and caught additional errors. Sandy Segal's knowledge of the European mathematical scene in this century is truly staggering. The readers suggested that a sampling of papers of Pólya's be reprinted as appendices, an idea I liked very much. You will note inclusion of four pieces from the *American Mathematical Monthly*. I'm very grateful for all of these contributions. I lay claim to any remaining errors in the text.

I also wish to thank the London Mathematical Society for graciously allowing certain materials here to be reprinted from the 1987 obituary prepared by Lester H. Lange and myself for the *Bulletin of the London Mathematical Society*. This includes essays on specific aspects of Pólya's work by Ralph P. Boas, Kai Lai Chung, D. H. Lehmer, R. C. Read, Doris Schattschneider, M. M. Schiffer, and Alan Schoenfeld, as well as a bibliography of Pólya's work, now brought up-to-date for the present volume. And for help in preparing the manuscript I relied on Mary P. Jackson and, for the picture reproduction, Dave Jackson and Leonard Klosinski.

It is always a pleasure to acknowledge the extraordinary skills and helpfulness of the publications staff of the MAA, Beverly Ruedi and Elaine Pedreira Sullivan, in this case. They are unfailingly efficient and professional, but always cheerful even when faced with vacillating and inefficient writers like myself. And I always rely on the good advice of my longtime colleague, Don Albers, who keeps the MAA publications program right on course.

G. L. Alexanderson
Santa Clara, California
June, 1999

Notes on References and Mathematical Commentary

There are references throughout the text to publications of Pólya's and those of others. A reference of the form [date, first number, second number] is to a publication of Pólya's and appears in the Bibliography; the first number refers to the number of the publication for that year, the second number being a page number in the case of a book. A reference of the form (name date; number) is to something in the list of references and gives author and date of publication, with a number following giving a page number, but only in the case of books.

Since not all readers will be interested in some of the more technical and mathematical comments, these sections are generally set apart in the text and may be passed over without interrupting the narrative.

Table of Contents

Prologue

Pólya and Euler

The similarity between Pólya's taste in problems and Euler's may have prompted Donald E. Knuth's remarks at Pólya's ninetieth birthday party at Stanford in 1977: "A few years ago when I was visiting the home of Professor [N. G.] de Bruijn in Holland, he and I asked ourselves the question:

> If we could be any mathematician in the history of the world (besides ourselves), who would we rather be? After some discussion we narrowed the choices down to Euler and Pólya, and finally settled on George Pólya because of the sheer enjoyment of mathematics that he has conveyed by so many examples.

Later de Bruijn elaborated on this idea with remarks in his retirement address from the Technological University Eindhoven, when he said:

> A mathematician who possibly more than anyone else has given direction to my own mathematical activity, is G. Pólya. All his work radiates the cheerfulness of his personality. Wonderful taste, crystal clear methodology, simple means, powerful results. If I would be asked whether I could name just one mathematician who I would have liked to be myself, I have my answer ready at once: Pólya. (de Bruijn 1985)

The influence of Pólya on these first-class mathematicians was all the more remarkable because mid-twentieth century mathematics in the United States was dominated by the influence of Bourbaki, that school of mainly French mathematicians who wrote under the pseudonym, Nicolas Bourbaki, and who set a very abstract tone for mathematics that continues to influence the direction of mathematics right into the 21st century. It was the time of the structuralists—the followers of the Polish, French and Russian schools of topology and the successors to the Italian

Leonhard Euler

algebraic geometers. But a steady interest in classical analysis persisted, with concentrations of analysts at New York University and Stanford University, among others. One of the giants in classical analysis of this century was George Pólya (1887–1985), who had come from that extraordinary group of mathematicians in Hungary and passed through Göttingen and Paris, to the Eidgenössische Technische Hochschule in Zürich and after detours to Oxford, Cambridge, Princeton, and Brown, eventually to Stanford.

Pólya was not a structuralist. He was interested in problems, making him, in a sense, a distant heir to Euler, who has been contrasted to another mathematician of his day, Lagrange. It's an oversimplification, but Euler dealt in problems, albeit magnificent problems, Lagrange in rigorous arguments and structure.

It should not be surprising that people have thought of Pólya and Euler as in some way related since Pólya himself had said without hesitation, when he was asked which mathematician of the past he admired most, "Euler." He went on: "Among old mathematicians, I was most influenced by Euler and mostly because Euler did something that no other great mathematician of his stature did. He explained how he found his results and I was deeply interested in that. It has to do with my interest in problem solving." (Albers and Alexanderson 1985, 251) He did admit that he was not familiar with all the work of Euler—who could be, given Euler's more than eighty folio volumes of published work? But a couple dozen volumes of Euler's *Opera Omnia* sat on Pólya's bookshelves and he found inspiration in them.

The Joy of Discovery

Pólya not only shared with Euler a taste for concrete problems and mathematics grounded in physical problems, he also shared with him a breadth of interests. Euler wrote on problems from analysis, calculus of variations, mechanics and other forms of applied mathematics, the theory of numbers, and questions that would eventually be considered parts of topology and graph theory. He also wrote on the theory of music, naval architecture, astronomy, as well as other areas. His *Lettres à une princesse d'Allemagne* covered all kinds of topics, including, for example, discussions of why the sky is blue and why some musical chords are more pleasing to the ear than others.

Similarly, Pólya's work ranged over various kinds of mathematics—analysis, of course, but also the theory of numbers, geometry, probability, combinatorics (using group theory) and graph theory. And there were even papers of his on voting schemes and astronomy. Further, he wrote extensively on heuristics and problem solving. When asked how he happened to have worked in so many different branches of mathematics, Pólya answered: "I was partly influenced by my teachers and by the mathematical fashion of that time. Later I was influenced by my interest in discovery. I looked at a few questions just to find out how you handle this kind of question." (Albers and Alexanderson 1985, 247)

So here is an example of a first-class research mathematician who was also respected and admired for his contributions to the teaching of mathematics through his many writings on problem solving and discovery. And his own research career was affected by this curiosity about how people solve problems.

Of course, his influence may have been less widespread had his personality been less ebullient. Pólya simply radiated enthusiasm and an almost child-like glee at discovering beautiful things in mathematics and in sharing his discoveries with others. It is small wonder that his students revered him. But for him, in both his research and his teaching, the joy of discovery was everything.

～ I ～

Childhood

Budapest

Budapest in the last decades of the nineteenth century was a sophisticated and lively place. It had not always been so. For centuries Hungary had been a country largely dependent on agriculture and Budapest had been a city of slaughterhouses and flour mills. In fact Budapest produced more flour than any other city in the world until 1900, when it lost that distinction to Minneapolis. In the period between the *Ausgleich* (Compromise) of 1867 and the First World War, however, there was a remarkable flowering of commercial and industrial development. Long subject to occupation by foreign powers—the Mongols, the Turks, the Austrians—Hungary became in 1867 an almost equal partner with Austria in the Austro-Hungarian Empire. The *Ausgleich* established the Dual Monarchy, *Kaiserliche und Königliche* (imperial and royal), combining Austria and Hungary. Hungary was the kingdom. Of course, the emperor-king was a Hapsburg and resided in Vienna. But this was the closest Hungary had come to independence in many years. The result was a period of relatively liberal politics and, at least in the urban areas, considerable prosperity.

The citizens of Budapest used this comparative freedom and financial well-being to develop institutions enhancing the artistic and intellectual life of the city. In earlier days there were stylish salons where ideas of the day could be openly and vigorously discussed. But by the late nineteenth century the preferred setting, at least for the less socially

Café New York

elevated yet intellectually curious, was the coffee house. (Lukács 1988, 148) These establishments sprang up all over the city and some—most notably the Café New York—survive to this day. The life of the universities extended right into these coffee houses.

It was during this period, between 1867 and roughly the turn of the century, that the great monuments of Budapest came to be built. The Parliament Building was begun in 1884 and completed in 1902. It is surely the most imposing edifice in the capital and something that has largely become the symbol of the city: what the Eiffel Tower is to Paris and the Houses of Parliament are to London. It is in the fashionably neo-Gothic style of the period. Construction on the State Opera House was begun in 1875 and upon completion in September of 1884, the well-known Hungarian composer, Ferenc Erkel, conducted its orchestra on opening night. It was Erkel who, in 1844, had written the national anthem and two of his operas on nationalistic themes are still performed, *Hunyadi László* and *Bánk Bán*. Somewhat earlier another famous Budapest institution had been built, one more relevant to the present account: the Hungarian Academy of Sciences (Magyar Tudományos Akadémia). This building was completed in 1864 and still serves as a reminder of the important role that science has played in Hungarian intellectual life. The principal façade contains statues representing the original six parts

Parliament, Budapest

of the Academy: history, law, linguistics, mathematics, philosophy, and sciences. There are also statues of six eminent scientists, four of whom can be considered mathematicians: Descartes, Galileo, Leibniz, and Newton. Among the six, only one is Hungarian: the linguist Révay. (The sixth is Mikhail Lomonosov, the Russian linguist.) (Dent 1990, 176, 182, 196)

The Hungarian Academy of Sciences

So, although Budapest did not have the beautiful ancient quarters that some of its rival European capitals possessed, by the late nineteenth century it had developed into a city with grand monuments and an impressive city plan. It is interesting to note that in 1901 a commission of the United States government, when assigned the task of studying a redesign of Washington, DC, planned a visit to six of what were considered to be the most important and well-designed cities of Europe. Budapest was among the six. The visitors included such renowned American city planners as Daniel H. Burnham (proponent of the City Beautiful concept), Charles F. McKim (of McKim, Mead, and White), and Frederick Law Olmsted (responsible for New York's Central Park and the campus of Stanford University). (Lukács 1988, 48)

Being Hungarian and Jewish

The Jewish population of Hungary made up about five percent of the population, though it made up over 23% of the population of Budapest. By 1910, 53% of those engaged in industry, 64% in trade and finance, 59% of medical personnel, and 62% of those practicing law in Budapest were Jewish. (McCagg 1972, 30) And contrary to the experience of Jews in nearby countries, the Jewish population of Budapest was subjected to little open anti-Semitism, at least until after the turn of the century. In fact there was considerable immigration of Jews into Hungary from Poland and other eastern European countries. In 1867 Jews had been given full civil rights and many even joined the ranks of the nobility. Twenty-eight families entered the peerage with baronial rank; seventeen acquired hereditary or personal membership in the Upper House of Parliament; and ten became Privy Counselors. In this sense Hungary was more liberal than many of its neighbors, so the climate for Jewish families was relatively good. There was anti-Semitism, but it remained largely below the surface. (McCagg 1972, 22)

In his biography of John von Neumann, Macrae addresses the intellectual advances in Hungary beginning in the latter part of the nineteenth century as follows: "The answer to 'the huge mystery of 1890–1930 Hungary's educational achievement' lies largely in . . . Hungary's post-1870 willingness to import middle-class Jews—because of its Magyar aristocracy's contempt for its majority of non-Magyar peasants—attracted the brightest and most educationally ambitious of Europe's Jews to Budapest . . . a cultured and upwardly mobile group, intent on giving their sons (sadly, more than their daughters) the education that some of them had never had. Many arrived in the 1890s when Hungary's counts and barons and monks and pastors had a new ambition: to produce more brilliant scholars and generally cultured young men than Vienna did. A discriminately excellent supply streamed into a competitively excellent schools system." (Macrae 1992, 64–65)

Jewish families tended to progress through the various social strata in a predictable way. Many started out in trade, moving in later generations into the professions—medicine and law—and eventually ending up with at least some family members in the academic world, the world of scholarship. The goal of achieving an academic appointment was consistent with the long-held respect for learning in Jewish culture.

The Family

György (George) Pólya was born on December 13, 1887. Of Jewish descent, his grandfather on his father's side, Mózes Pollák, had been a minor merchant. The surname Pollák indicates possible Polish origins, but the early history of the family is not well-documented. The family may have come from Poland, Lithuania, or the Ukraine. George's father, Jakab Pollák, who was born in Békésszentandrás on the great plain of Hungary in 1844, thought first of becoming a doctor. But he could not stand the sight of blood—recognized as not a promising trait in a medical student—so he moved instead into the law. One of ten children, he had had to overcome financial hardship to get an education and upon finishing he went to work for a solicitor as a junior partner. When the senior partner left the firm to go into business, Jakab took over the practice. (J. Pólya 1986)

George Pólya's mother, Anna (née Deutsch), was born in 1853 to a family that had probably moved farther along on the social scale than her husband's. The maiden name of Anna's mother was Blau and this family claimed residence in Buda at the time of the Turkish occupation (sixteenth and seventeenth centuries). By the eighteenth century they were fairly well-to-do. On Anna's father's side the family could be traced to the small towns of Arad and Lugos, now part of Romania. By the late nineteenth century members of this family were prosperous—with money made in textiles—and had moved into the professional and intellectual classes. A cousin, Ágoston Kanitz (1843–1896), was a university professor of botany, and descendants of his sister included two doctors and a general. (J. Pólya 1986)

George Pólya at an early age

Unfortunately Jakab's law practice was not successful in spite of his "brilliance and honesty." According to family tradition, he threw "ink bottles at clients seeking legal excuses for ethically indefensible purposes." (J. Pólya 1986) After the failure of the practice a distant relative on his mother's side arranged something with his own employer so that Jakab could move into a position with an insurance company as a legal officer. The company was the Budapest office of one of the leading insurance companies in the Austro-Hungarian Empire, the Assicurazioni Generali of Trieste. This position provided an income for the family but an income probably insufficient, at least in the eyes of Anna. She may have felt that she had married beneath her station. In any case there was pressure on Jakab to improve his standing in the community. The pressure was partly self-generated, evident in his determination to move into the academic world, but partly it came from his awareness of the family's need for money.

The Pólya Technique

The couple had a total of six children, though one daughter died shortly after birth. The oldest brother, Jenő (Eugene), loved mathematics, but, to

Family picture, ca. 1889

his later regret, he studied medicine instead. He went on to become an eminent surgeon, one of the most renowned in Hungary and one with an international reputation derived from his developing the Pólya technique of gastrectomy, a form of stomach surgery recognized and advocated by the American surgeon, William G. Mayo, among others. In some circles the name Pólya prompts recollections of Jenő's contributions to surgery, not George's contributions to mathematics. Jenő was eleven years older than George. Between Jenő and George were two sisters, Ilona, ten years older, and Flóra, eight years older. Then there was another brother, László, who was four years younger, but he was killed in Galicia during the First World War. Of the three brothers, the family is reported to have considered László the most intellectually gifted. (Sandor and Modlin 1995)

To support this family Jakab worked long hours during the day at the insurance company and then worked into the evening writing various scientific treatises, largely in search of a mathematical understanding of economic fluctuations. Unfortunately the mathematical techniques for this were not available to him and he was frustrated in his efforts. He did, however, produce ten books and numerous tracts, some to convince poverty-stricken Hungarian peasants of the advantages of cooperative techniques practiced by farmers in the West. And he translated into Hungarian Adam Smith's *An Inquiry into the Nature and Causes of the Wealth of Nations*. The translation was used as a university text for several decades.

Jakab's aim through all of this was to obtain a position at the university and he did achieve some level of success in this. He was named lecturer at the Péter Pázmány University of Budapest and, shortly before his death, was elected to the Royal Hungarian Academy. Nevertheless, probably as a result of overwork, in combination with other factors, he died at the age of 53 when George was ten years old. This put even more of a financial burden on the family. The two sisters promptly went to work for the Assicurazioni Generali. Ilona never married and worked at the insurance firm throughout her lifetime, but Flóra did marry and her elder son, Sándor Kemény, became a mathematician. He was killed by the Nazis during the Second World War. His widow, Katalin (Kati) Pataki, remarried but remained close to the family. In later years she translated several of George Pólya's mathematical works into Hungarian.

After his father's death, Jenő had to assume additional responsibility for the family. He was to marry and have four children, János (John), Mihály (Michael), Zsuzsi (Suzanna or Susi) and Juci (Judi), all of whom moved eventually to Australia except Susi who lives in Oberlin, Ohio. Jenő received many honors in his lifetime, including honorary fellowship in the American College of Surgeons. This honor came in 1939 and five years later he was killed in Budapest by the Nazis. (Sandor and Modlin 1995)

Jakab had the reputation for being a man of integrity and of great scholarship and industry, but he had little time to spend with his family. As early as 1882 he changed his name from the family name Pollák to a rustic Hungarian name, Pólya, probably to make the name sound less Jewish and hence to enhance his chances for an academic position. (This was not

Jenő Pólya as a young man

uncommon. The family name of the famous scientist-philosopher, Michael Polányi, for example, had been Pollacsek up to eight years following Polányi's birth. The practice resulted too from the pressures from Hungarian nationalists to convince non-Magyars to take Magyar names.)

Though both of George's parents were Jewish, his mother's grandfather had converted to Christianity and had become a Presbyterian. But Jakab in 1886 converted to Roman Catholicism and he, his wife, and the three children at that time were baptized. The younger children, George and László, were baptized in the Catholic Church shortly after birth. Ever concerned about social respectability, George's father arranged to have as one of the witnesses at George's baptism a director of the insurance company where he worked, a nobleman of an old and respected family. So George's Godfather was Count Sándor Károlyi, Inner Secret Councilor, who gave his occupation on George's birth certificate as "large estate owner." The Károlyi family was one of the best-known and richest in Hungary, among the top five of the High Aristocracy, having worked out with the Hapsburgs the modus vivendi after the surrender to Austria in 1711. Even the priest presiding at the christening, Dr. György Kanyurszky, identified himself as "university professor." It was an auspicious beginning.

George Pólya's birth certificate

~ 2 ~

Pólya's Education

Around the turn of the century, Hungarian schools and the Hungarian university system were organized along German lines. This is not surprising since this style prevailed throughout much of Europe and the ideal of the German university was influential even in the United States. The upper level literary Gymnasium required both Latin and Greek, and a good performance in such a Gymnasium assured one a place at the university. Among the scientists produced by Hungary in the late nineteenth century and early twentieth century, some attended the Royal Joseph University of Technology (Müegyetem): Tódor (Theodore) von Kármán and Leó Szilárd, among others. More, however, attended the University of Budapest, and this list included George Pólya and János (John) von Neumann. By the time Pólya entered the University there were few if any restrictions on the education of the Jewish population. In the early part of the nineteenth century, University positions had been available to Jews only if they planned to study medicine. (McCagg 1972)

Pólya began his education in a local elementary school, where he received a certificate of excellence for "diligence and good behavior" in 1894. The family was at that time living at Izabella-u. 81, Budapest, in the Theresia District a few blocks from the West Railroad Station and from the Liszt Academy of Music (Liszt Ferenc Zeneakadémia). Shortly after Jakab's death the family, now headed by Anna and the eldest son, Jenő, moved to the more central Leopold District, to Arany János-u. 29, a street running perpendicular to the Danube and just back of the Academy of Sciences. The location was in the center of Pest and within walking distance of the Gymnasium Pólya attended on Markó-u., named for the poet, Dániel Berzsenyi. Pólya's father and older brother had attended this Markó Street Gymnasium as well. (J. Pólya 1988) There Pólya studied Hungarian, German, Latin, and Greek, as well as other usual subjects of study at that level. Oddly enough, his grades in mathematics were not among the highest. In fact in three of four years of "geometrical drawing", he received a grade of "jó" (good) only once and an "elégséges" (satisfactory) in the other three. In arithmetic or "calculating" he did better—a "jeles" (excellent) for four years, "jó" for three years and "elégséges" for one. This was in contrast to the high grades, "jeles" and "kitűnő" (outstanding), he was receiving in languages, geography and other subjects. (Of course, the grading system in the Gymnasium may not have been entirely reliable—the Nobel Prize-winning biochemist, Albert Szent-Györgyi, once received a "C" in physics, and Béla Bartók a "B" in composition!) (Lukács 1988, 146) Overall Pólya's record at the Gymnasium was a very good one, but he was not a "straight A" student. And, if anything, he did show more aptitude for languages and the areas that we would today classify as the humanities, than he did for the sciences.

The elementary schools and the curriculum of the Gymnasium heavily favored memorization. This may not have been the best program for a budding mathematician or scientist, but it did leave Pólya with an impressive fund of quotable material from the classics. Well into his 80s he could recite at great length from Homer's *Iliad*, from the beginning, in Greek. And his command of Dante's *La Divina Commedia* in the original was impressive. Of course, he could also quote at length from Schiller, Goethe and, his favorite, Heinrich Heine.

"Mathematics Is in Between"

Given the financial problems that the family had endured after the death of Pólya's father, his mother was anxious that he complete his studies. She wanted him to study the law and follow in his father's footsteps, so he tried this for a while but found it boring. In fact he lasted only one semester as a law student. He then shifted into the study of languages and literature for two years. His transcripts at the University for this period show a large number of courses in literature and linguistics in various languages, and at the end of the two years he passed an examination that resulted in his receiving a teaching certificate for Latin and

Pólya during his student years in Budapest

Hungarian so that he could teach in the lower classes of the Gymnasium, children between the ages of ten and fourteen. He never made use of this teaching certificate, but he was surprisingly proud that he had obtained it and he usually mentioned it in conversations dealing with that period of his life.

When he was taking courses in philosophy he fell under the influence of Professor Bernát Alexander, the grandfather of the eminent Hungarian mathematician, Alfréd Rényi. Alexander encouraged him to take some courses in physics and mathematics. Among physics, philosophy, and mathematics he was probably least interested in the courses in mathematics. This may have prompted him to explain in what has now become quite a well-known quote: "I thought 'I am not good enough for physics and I am too good for philosophy. Mathematics is in between.'" (Albers and Alexanderson 1985, 248)

Of course, he was indeed good enough to do physics and his dismissal of philosophy was made largely because it made a clever, epigrammatic statement. He was, however, on many other occasions somewhat disdainful of philosophy. He was fond of applying to philosophers a variation of the frequently heard French description of graduates of the *Hautes Écoles* in France by saying that "philosophers know everything but not much else."

His seeming reluctance to move into mathematics may have been partially the result of his experience in the Gymnasium, where he gave high marks to his teachers in Latin, Hungarian, and geography, but of his three mathematics teachers there he was later to remark that "two were despicable and one was good." (Wieschenberg 1987)

At that time there were a number of school competitions in Hungary. Pólya received a prize for his translation of Heine into Hungarian, for example, but it is reported that when he took the famous mathematical examination, the Eötvös Competition, he failed to turn in his test paper. Wieschenberg wrote of this from interviews she had with Pólya in 1983 and 1985. McCagg claims that Pólya was among the winners of the Eötvös, along with Ede (Edward) Teller, Leó Szilárd, and John von Neumann. (McCagg 1972, 214-215) This claim is not borne out, however, by the report of winners in Kürschák's collections of Eötvös problems (Kürschák 1963). There is little doubt, however, that the Eötvös Competition was influential in motivating many Hungarian students at the time. Pólya's well-known co-author, Gábor Szegő, won the Eötvös in 1912; the extensive list of other winners includes such illustrious names as Lipót Fejér, Dénes Kőnig, Alfréd Haar, Marcel Riesz, and Tibor Radó. Since 1947 the contest has been known as the Kürschák Competition since the Eötvös name was taken over by the physicists for a physics competition. The list of names of distinguished winners since 1947 is still impressive; it includes Béla Bollobás, László Lovász, László Babai, and János Kollár. (Császár 1997)

Fejér

One of the other factors almost certainly responsible for the flowering of Hungarian mathematics around the turn of the century was the influence of a well-known mathematician and professor at the University, Lipót Fejér. Here was a professor of mathematics who gave brilliant lectures but also exerted an influence on talented students far beyond their experience in the classroom. He had a charismatic personality and great style. Pólya readily admitted that he had personally been much influenced by Fejér, and mathematically he had, in turn, influenced a couple of Fejér's papers. Further, Fejér had influenced a couple of his, though Pólya's taste in problems had been far more influenced by others. (Albers and Alexanderson 1985, 250)

Pólya was to cite one other factor in the unusual flowering of mathematics at that time: a journal that posed challenging problems to students. Also when asked why Hungary should have produced so many mathematicians and not similar numbers of world-class figures in other disciplines, Pólya pointed out that Hungary is not a rich country and mathematics, unlike disciplines such as chemistry and physics, does not require vast resources for equipment. This argument is not entirely convincing, however, since Hungary not only produced mathematicians of the first rank, but also physicists. The list of physicists includes, in addition to the venerated Loránd Eötvös, Szilárd, Teller, Dénes Gábor, and Jenő (Eugene) Wigner, an impressive list. The list of internationally recognized mathematicians is, however, even more impressive and certainly much longer. One can argue about who is included and who is not, but the list given by Pekonen in 1991 is certainly enough to convince one that something right was happening mathematically in Hungary in the latter part of the nineteenth century and the early part of the twentieth: József Kürschák (1864–1933), Frigyes Riesz (1880–1956), Marcel Riesz (1886–1969), Lipót Fejér (1880–1959), Ottó Szász (1884–1952), Alfréd Haar (1885–1933), Mihály Fekete (1886–1957), György Pólya (1887–1985), Tibor Radó (1895–1965), Gábor Szegő (1895–1985), Béla von Kerékjártó (1898–1946), János von Neumann (1903–1957), Arthur Erdélyi (1908–1977), Pál Turán (1911–1976), Pál Erdős (1913–1996), Alfréd Rényi (1921–1976), and Imre Lakatos (1922–1974). (Pekonen 1991, 190–191) There had been earlier Hungarian mathematicians of note, of course, the most important being Farkas and János Bolyai, father and son, as well as the set theorist, Gyula Kőnig (1849–1913), whose son, Dénes Kőnig (1884–1944), is often called the father of graph theory, and Ludwig Schlesinger, sometimes assumed to be German since he spent most of his career in Germany, the first Hungarian mathematician to hold a professorship there. But the golden age of Hungarian mathematics, at least up to this time, was the period between the latter part of the nineteenth century and the middle of the twentieth. There were three dominant themes: classical analysis inspired by the work and influence of Fejér, linear functional analysis, largely inspired by Frigyes Riesz, and the combinatorial mathematics of Erdős and Turán. Each of these schools had influence far beyond the borders of Hungary. (Hersh and John-Steiner, 1993)

It is hard to think of another discipline where the Hungarian influence has been so strong. Hungarian painting and sculpture are essentially unknown outside Hungary. Partly due, probably, to the inaccessibility of the Hungarian language, Hungarian literature is little known outside the country except possibly for the plays of Ferenc Molnár, popular internationally in translation. A Hungarian composer of international repute was Ferenc (Franz) Liszt, with, more recently, Béla Bartók, Zoltán Kodály, and, perhaps, György Ligeti, Miklos Rósza, and Ernst von Dohnányi, enjoying international recognition. In a lighter vein there are Ferenc (Franz) Lehár and Imre (Emmerich) Kálmán. A Hungarian composer highly thought of in Hungary in earlier times was the composer of operas, Ferenc Erkel, mentioned earlier, but his work is largely ignored outside Hungary. In conducting the strong presence of Hungarians is clear: Antal Dorati, Eugene Ormandy, Fritz Reiner, George Solti, and George Szell. In film direction there are Alexander Korda and Michael Curtiz (né Kertész), the director of Casablanca, and in film acting, Béla Lugosi, Peter Lorre, Leslie Howard, Paul Lukas, and Tony Curtis. In architecture, there are Marcel Breuer and László Moholy-Nagy (interesting, incidentally, for the mathematical ideas in his work for the Bauhaus). And so it goes. These people were influential in their fields, but in each field the number is small and the contributors are normally not of the absolutely first rank. But in the sciences, and in particular in mathematics, many have been first class and the Hungarian influence has been widely felt.

Let us go back and examine Pólya's three major factors in explaining this Hungarian mathematical renaissance: the journal, the competition, and Fejér. The journal, the *Középiskolai Matematikai Lapok*, was published for secondary school students and was in a sense tied to the competition, in that doing the problems posed in the journal provided the practice for competitions to come. The competition itself, named for the eminent physicist Loránd (Roland) Eötvös, son of the great Hungarian patriot, author, and political figure, Baron József Eötvös, was established in 1894. It was open to young people who had passed the terminal examinations of the Gymnasium and were ready to enter the university. The competition is still being administered in Hungary, though, as mentioned earlier, it is now named for Kürschák. In some ways it provided the model for many other similar secondary school level competitions, which now culminate in the International Mathematical Olympiad.

What about the influence of Fejér? Lipót (Leopold) Fejér was only seven years older than Pólya, born in 1880 in provincial Pécs. He took only a second prize in the Eötvös Competition, but he started his university studies in Budapest and completed his Ph.D. there at the age of 22. Prior to finishing the Ph.D. he spent an academic year, 1899-1900, at the University of Berlin where he was enormously influenced by Hermann Amandus Schwarz. It was a suggestion of Schwarz's that led to what is referred to as Fejér's theorem and provided the topic for his Ph.D. dissertation. The theorem states in essence that a Fourier series is summable (C, 1) to the value of the function for each point of continuity. This has a number of important corollaries including the Weierstraß Approximation Theorem; other consequences of this theorem were pointed out by Pólya in an article on Fejér. [1961, 4] Fejér rose through the academic ranks quickly and was a full professor at the University of Budapest at the age of thirty-one, one year prior to Pólya's receiving his Ph.D. at the University.

"Half Eccentric"

In trying to explain Fejér's appeal to younger mathematicians, Pólya once raised the question of whether Fejér was "eccentric". He answered by saying, "If you could see him in his rather Bohemian attire (which was, I suspect, carefully chosen) you would find him very eccentric. Yet he would not appear so in his natural habitat, in a certain section of Budapest middle-class society, many members of which had the same manners, if not quite the same mannerisms, as Fejér—there he would appear about half eccentric." [1969, 4] In a famous picture of Fejér [1987, 2: 39] he is shown strolling down a fashionable street with a homburg and cane, looking very much like a continental boulevardier.

Fejér was remembered not only for his mathematics but for his witty and unexpected remarks as well. Pólya provided an example of an occasion at the meeting of the Deutsche Mathematiker-Vereinigung in Leipzig in 1922. Pólya's wife was an avid, though amateur, photographer at that time and Pólya was an underpaid *Privatdozent* in Zürich, who had not yet

Lipót Fejér

been able to obtain an appointment as a regular professor. As Pólya put it, "She ... stopped Fejér in the company of three or four [other mathematicians], in front of the University on the streetcar tracks, took a picture and was about to take a second one as Fejér spoke up. 'What a good wife! She puts all these full professors on the tracks of the streetcar so that they may be run over and then her husband will get a job!'" [1987, 2: 58]

Pólya wrote the following about Fejér's remarkable personality:

"He had artistic tastes. He deeply loved music and was a good pianist. He liked a well-turned phrase. 'As to earning a living', he said, 'a professor's salary is a necessary, but not a sufficient, condition.' Once he was very angry with a colleague who happened to be a topologist, and explaining the case at length he wound up by declaring: '... and what he is saying is a topological mapping of the

F. Riesz, Rademacher, Szász, Knopp, Mrs. Szegő, Bessel-Hagen, and Ostrowski, all on the streetcar tracks

truth.' He had a quick eye for human foibles and miseries; in seemingly dull situations he noticed points that were unexpectedly funny or unexpectedly pathetic. He carefully cultivated his talent of raconteur; when he told, with his characteristic little gestures, of the little shortcomings of a certain great mathematician, he was irresistible. The hours spent in continental coffee houses with Fejér discussing mathematics and telling stories are a cherished recollection for many of us. Fejér presented his mathematical remarks with the same verve as his stories, and this may have helped him in winning the lasting interest of so many younger men in his problems." [1961, 4]

Those who knew Pólya will recognize in this some of the influence that Fejér must have had on Pólya himself. Pólya too was never happier than when he could tell stories about mathematics and mathematicians. Much like Fejér he had a taste for jokes, for bon mots, for epigrammatic remarks.

Pólya probably was also influenced by Fejér's style in writing and explaining mathematics. Again, in 1961, Pólya said,

Fejér talked of a paper he was about to write up. 'When I write a paper,' he said, 'I have to rederive for myself the rules of differentiation and sometimes even the commutative law of multiplication.' These words stuck in my memory and years later I came to think that they expressed an essential aspect of Fejér's mathematical talent: his love for the intuitively clear detail.

It was not given to him to solve very difficult problems or to build vast conceptual structures. Yet he could perceive the significance, the beauty, and the promise of a rather concrete, not too large problem, foresee the possibility of a solution, and work at it with intensity. And, when he had found the solution, he kept on working at it with loving care, till each detail became fully intuitive and the connection of the details in a well-ordered whole fully transparent.

It is due to such care spent on the elaboration of the solution that Fejér's papers are very clearly written, and easy to read and most of his proofs appear very clear and simple. Yet only the very naive may think that it is easy to write a paper that is easy to read, or that it is a simple thing to point out a significant problem that is capable of a simple solution. [1961, 4]

Pólya, again like Fejér, tried to explain ideas with the greatest care, writing and rewriting, always looking for a way to make the idea absolutely clear to the reader.

~ 3 ~

Vienna, Göttingen, and Paris

Singing the Wrong Songs

Pólya's university years began with intensive studies of Hungarian, Latin, linguistics, and philosophy. He stuck close to his studies except for an occasional foray into politics. The government in Hungary, though relatively liberal, nevertheless probably viewed students the way many governments do, as potential threats to the maintenance of order. Pólya would occasionally describe a run-in he had once with the police, when he and a group of other students were on a riverboat and persisted in singing songs that the police deemed revolutionary.

Pólya was also one of the founding members of the famous *Galilei Kör* (Galileo Circle) at the University. Karl Polányi, who was the brother of the more famous Michael Polányi (the biochemist-physicist-mathematician-social scientist-philosopher), was the first president of the group. From a modern vantage point this band of "radical intellectuals" would not appear to be very subversive. They sponsored lectures, published a journal for young scholars, and opened reading rooms. The slogan of one of the Circle's mentors, Professor Julius Pikler, was: "Do not put religious, social, or political considerations above scientific research and speculation." Pólya gave a lecture to the group on one of his heroes, Ernst Mach, so the content of the programs was by no means totally political. Nevertheless, the members did see themselves as socialists and there were lectures on Marx and others who certainly would not have been approved of by the authorities. Pólya in later years, however, re-called the meetings as being dull and innocuous. In spite of that, their activities were sufficiently threatening to the government that the Circle was ordered dissolved in early 1918 because of its subversive propaganda. And there appears to have been a direct connection between the Circle and the revolution of 1919. This was, to be sure, well after Pólya had left the university and even after he had left Hungary.

Vienna—The Magnet

While at the University of Budapest he enjoyed the benefit of free tuition. Additional expenses were covered by his older brother Jenő who, by that time, was a practicing surgeon. In the year prior to his receiving the Ph.D. in 1912, however, Pólya decided to spend a year at the University of Vienna. As lively as Budapest was, and even though he could hear lectures in Budapest by Fejér in mathematics and Loránd Eötvös in physics, Vienna nevertheless was a magnet for anyone in the Austro-Hungarian Empire and indeed others as well. The University of Vienna was not in a class with Paris or Berlin or Göttingen in mathematics, but it did have a few figures of some renown.

In order to pay expenses, Pólya took on the task of tutoring the rather dim son of a very rich nobleman in Vienna. He found the work difficult—the student was, according to Pólya's comments in later years, extraordinarily dull—but it did give him the opportunity to attend several seminars by Wilhelm Wirtinger and to take classes in algebra from Franz Mertens. Mertens, of course, had done important work in the theory of numbers and in algebraic geometry while he was in Graz, and by the academic year 1910–11 he had moved to Vienna. According to Pólya's *Meldungsbuch* (report book) from the University these are the two mathematicians with whom he had direct contact in Vienna. In addition, he attended lectures on a variety of topics in physics—including relativity and optics, among others.

The Vienna experience seems not to have had a major influence on his mathematical tastes. The contact with Mertens stimulated no great interest in algebra, a subject for which Pólya seemed to have little sympathy. Furthermore he found the city of Vienna less appealing than Budapest. The government was more repressive than in Hungary and there was a higher level of anti-Semitism, so he was happy to have an opportunity to return to Budapest immediately after the close of the academic year in Vienna. He proceeded to complete his dissertation, which he wrote on some problems of geometric probability with associated integrals. He wrote it with little or no help from faculty since there was no one on the faculty in Budapest in this field. The type of geometric probability that was used is no longer fashionable and though the dissertation was later published in the *Mathematische Annalen*, this work is not really related to his later major contributions in the area of probability theory. Some have suggested that he wrote his dissertation in the field of probability because of the longtime association of his family with insurance. It's a possibility, but there is no record in Pólya's papers to indicate any such connection.

Göttingen

Following Pólya's completion of the Ph.D. in Budapest he elected to spend two postdoctoral years at the University of Göttingen. If Vienna had been disappointing, Göttingen was quite the opposite. This was the university of Gauss, Riemann and Dirichlet. Pólya did not attend lectures by Felix Klein—he had just retired—but he certainly was in contact with him while he was there. Many years later when Jean Pedersen was discussing with Pólya the connection between polyhedra and groups, she asked him where he learned this material. Pólya replied, "I learned it from the master—Felix Klein—in circumstances just like this." Pedersen remarked that the experience gave her a sense of the continuity of mathematics.

Pólya's *Anmeldungs-Buch* for his time in Göttingen reads like a "who's who" of early twentieth century mathematics. He attended classes offered by David Hilbert: courses in the theory of partial differential equations, the mathematical foundations of physics, and the elements of the foundations of mathematics, as well as two seminars in mathematical physics. In addition to those, he attended a course by Edmund Landau on infinite series and Fourier series, along with a course in mathematical problems. From Otto Toeplitz he took a course in the theory of invariants, and from Hermann Weyl a course on integral equations and their application in mathematical physics, as well as a course in "elliptic, abelian, and automorphic transcendentals." Constantin Carathéodory and Landau offered a course in set theory and another course in mathematical

A page from Pólya's Anmeldungs-Buch *at Göttingen*

problems, and from Carathéodory alone Pólya took courses in the calculus of variations and conformal mapping. From Erich Hecke he took a course in the theory of algebraic number fields. In addition to these courses, as he had done in Vienna, he took a series of courses in mathematical physics, one, a seminar, offered jointly by Carl Runge and Ludwig Prandtl, another by H. T. Simon, one by Max Born on energy, and a course in vector analysis by Woldemar Voigt. Some of the latter names are perhaps not so well known in the mathematical community, but they were well known to other scientists and to applied mathematicians. Voigt was a physicist at Göttingen. Simon was head of the Institute for Applied Electricity; Prandtl was associated with the Institute for Applied Mechanics and was, some would say, the founder of the science of aerodynamics. (Reid 1970, 97)

No place in the world at that time, with the possible exception of Paris, could rival Göttingen for mathematical power. Pólya seems to have

Edmund Landau

taken advantage of the setting by rubbing shoulders with the giants. Of course, the problems in classical analysis that he attacked over the next few years were no doubt problems that were being discussed in the Göttingen climate. It appears, however, that no one exerted any extraordinary personal influence over him in that period. One of his first major papers was coauthored with Mihály Fekete, but this was a connection from Budapest.

David Hilbert

Göttingen did however provide Pólya with some fine stories that he could pass along to his students, including stories about Hilbert. Famous or popular professors often inspire stories, probably not all true, but they become part of the standard lore. Pólya told the story about Hilbert's alleged absentmindedness. At a party that the Hilberts were giving at their house, Frau Hilbert noticed that Professor Hilbert had forgotten to put on a fresh shirt, so she asked him to go upstairs and put on a clean one. Obediently he went upstairs and after some time when he failed to appear, his wife went upstairs to investigate and found that he had gone to bed. It was the natural sequence of things: he took off his coat, then his tie, then his shirt, and went to bed. [1969, 4]

Another account of Hilbert that Pólya gave, he particularly liked because it reminded him of the Göttingen that he knew, a rather formal place with prescribed customs, particularly for young faculty. A new member of the faculty was obliged to put on a black coat and top hat, take a taxi around to the homes of senior faculty, and introduce himself in this way to his colleagues. He would present a visiting card at the door. If the senior faculty member was at home, then he was to go inside and chat for a few minutes and depart. One such colleague came to Hilbert's house and either Hilbert or his wife decided that he was "at home," so the young man came in, sat down, put his top hat on the floor, and proceeded to talk. So far, so good. The problem was that he did not stop talking. Hilbert, who was probably thinking about some mathematics at the time, became more and more impatient, so he finally stood up, took the top hat from the floor, put it on his own head, and said, "I think, my dear, that we have delayed the Herr Kollege long enough." And he walked out of his own house. [1969, 4]

One could compile a whole collection of stories about the Riemann hypothesis.[1] Pólya's story about Hilbert and the Riemann hypothesis goes as follows, in Pólya's words: "There is a German legend about Barbarossa, the emperor Frederick I. The common people of Germany liked him and as he died in a crusade and was buried in a faraway grave, the legend sprang up that he was still alive, asleep in a cavern of the Kyffhäuser Mountain, but would awake and come out, even after hundreds of years, when Germany needed him. Someone allegedly asked Hilbert, 'If you would revive, like Barbarossa, after five hundred years, what would you do?' 'I would ask,' said Hilbert, 'has somebody proved the Riemann hypothesis?'" [1969, 4] As of this date (1999) the conjecture remains unproved, one of the few remaining unsolved problems from the famous list of problems for the 20th century proposed by Hilbert in his address to the Paris International Congress in 1900.

---

[1] The Riemann hypothesis states that the nontrivial zeros of the Riemann zeta function, initially defined by $\zeta(z) = \sum_{n=1}^{\infty} 1/n^z$ for Re $z > 1$ and analytically continued into the whole plane, all lie on the line $x = 1/2$.

David Hilbert

Pólya was also fond of telling a story about Richard Dedekind, a student much earlier at Göttingen—he had studied there with Gauss—and a professor at the Polytechnion in Zürich before he went to Brunswick. Pólya told that when Dedekind retired, he lived in a quiet way, seeing very few people. The *Jahresbericht* of the German Mathematical Society reported that Dedekind had died, giving the day, the month and the year. Dedekind was, however, very much alive, so he wrote a letter to the editor with the text, "In your communication, page so and so, of the *Jahresbericht*, concerning the date, at least the year is wrong." Pólya saw this as a "sharp mathematical statement", one that is even sharper in German where one would use *"Jahreszahl"* (year number) for "year." (G. Pólya 1975)

Expulsion from Göttingen

Pólya's Göttingen experience, however, came to an unhappy end. On a train as Pólya was returning to Göttingen after holidays, he was struggling to put his suitcase on an overhead rack when it fell and struck a German undergraduate. Tempers flared and there was talk of a duel. This was reported to the University administrators and Pólya was in danger of being dismissed from the University. University documents describing all of this existed until shortly after Pólya's death, at which time his wife destroyed them to protect his reputation, not realizing that the incident was already fairly widely known. So accurate details of this contretemps are at this point lost. Pólya did, however, describe the event in a letter to Ludwig Bieberbach in 1921 when he was applying for a job. That letter survives. Here is how Pólya described the incident: "The story of the boxed ear is completely childish and, by the way, it goes like this: On Christmas 1913 I traveled by train from Zürich to Frankfurt/M and at that time I had a verbal exchange—about my basket that had fallen down—with a young man who sat across from me in the train compartment. I was in an overexcited state of mind and I provoked him. When he did not respond to my provocation, I boxed his ear. Later on it turned out that the young man was the son of a certain *Geheimrat;* he was a student, of all things, in Göttingen. After some misunderstandings, I was given a *consilium abeundi* (i.e., told to go away) by the Senate of the University. (The story, even privately, is not worth defending.)" So Pólya apparently decided not to contest the case. He abandoned Göttingen at that stage and he left almost immediately for Paris. (G. Pólya 1921)

Paris

It is an interesting topic for speculation, what might have happened had he stayed at Göttingen and taken the position as a *Privatdozent* in Frankfurt that had been offered to him. As it turns out he was probably better off not taking that position and instead taking the route that he did that led to Paris and Zürich. In any case it was an apt completion of his mathematical education that he went to Paris, the other major mathematical center of the day, where the legacy of Henri Poincaré was still felt. Hadamard was already at the Collège de France. And the French school was thriving: Émile Borel, Élie Cartan, Maurice Fréchet, Henri Lebesgue and the preeminent Émile Picard.

Pólya recalled his having been asked by some of his professors at Budapest to extend greetings on their behalf to the most important French mathematician at the time, Picard. He remembered having been very intimidated by Picard—terrified may not be too strong a term. Part of his concern was that he realized that he came from what would be viewed, at least in Paris, as a provincial environment, and he found the whole academic establishment in Paris intimidating. The actual encounter with Picard was not reassuring. He remembered him as a formidable and formal gentleman who was polite but gave the impression of not being very impressed by a postdoctoral student from Budapest. Picard later was to be critical in keeping German mathematicians from participating in international congresses following World War I. So his cool reception of Pólya could have been colored by the fact that Pólya had most recently been in Göttingen. The memory of the Franco-Prussian War lingered and World War I was imminent.

Pólya told the following story about Picard. Picard was, of course, a member of the Académie des Sciences, Paris, where he was later permanent secretary. Another member of the Académie, a professor of theoretical physics and mechanics, was Joseph Boussinesq. In Pólya's words, "Boussinesq had four wives, well understood not all at the same time, but one after the other. And he did not divorce these wives but they died one after the other. He survived all four wives, by what means the story doesn't tell. When a wife of Boussinesq died, then his colleagues at the Académie, at least the closest colleagues, had to be present at the funeral. At the funeral of the fourth wife the fellow academicians, two by two, walked through the cemetery. Picard went with a younger colleague, I think with Émile Borel. It started raining. Picard had no umbrella, so Borel offered his umbrella and his arm to Picard. Then Picard allegedly said: 'Thank you, my dear colleague. It is also very necessary. It rains so hard. I observe that it often rains at the funerals of the wives of M. Boussinesq.'" (G. Pólya 1975)

Though he was impressed by the richness of Paris life, in particular the intellectual life, he nevertheless seemed not to have enjoyed his time in Paris very much. He found the living quarters that were affordable to a foreign student with meager resources, little short of appalling. He had apparently never seen anything quite like the plumbing arrangements in the students' quarters and the images remained with him for seventy years. Life in Göttingen had been easier.

Lions, a Christian and Lebesgue

Still, there were advantages derived from his stay in Paris. He perfected his French, which quite possibly remained his favorite language. It prepared him for a lifetime of speaking French at home since his future wife was to be from the French-speaking part of Switzerland. And his encounter with Lebesgue provided him with his favorite opening to an after dinner speech. He heard this from Lebesgue: At the Colosseum in Rome a Christian was led into the arena and a ferocious lion was let loose to attack him for the delight of those assembled. The lion rushed up to the Christian, but the Christian whispered something in the lion's ear and the lion quietly crawled away. A second lion, even more ferocious, was sent into the arena and again the Christian whispered in the lion's ear, with the same result. After several such disappointing encounters, the Emperor told his deputy to tell the Christian he would be set free if only he would tell the Emperor what he was whispering to the lions. The Christian replied: "It's really quite simple. I just say, 'After dinner you have to give a speech.'"

Pólya stayed in Paris less than a year, enough time to make contacts with a number of mathematicians with whom he would later stay in close touch, but when he received an offer from the Eidgenössische Technische Hochschule (ETH) (the Swiss Federal Institute of Technology) in Zürich, an offer extended by Adolf Hurwitz, he accepted it eagerly. He moved to Zürich and was to remain there for twenty-six years, from 1914 to 1940.

~ 4 ~
Zürich

Hurwitz

With the probable collapse of the arrangements with the University of Frankfurt and with his general dissatisfaction with life in Paris, it is clear that by the winter of 1914 Pólya was looking for a job. A letter of application to Budapest survives, dated February 23, 1914. In this letter Pólya makes his best case for an appointment, outlining his various research results and publications at that stage. He also points out that his German is very good, his French is passable, and that he can read English and Italian. But the position that opened up to him was not in Budapest, it was the appointment as *Privatdozent* at the ETH in Zürich. He was extremely pleased to have an opportunity to go there because of the possibility of working with Adolf Hurwitz (1859–1919). In manner Hurwitz was as different from Fejér as it is possible to be. Fejér was colorful and Bohemian in appearance, perhaps almost flamboyant. By contrast, Hurwitz was very proper, conventional, respectable, and thoroughly middle class in his manner. At the same time there was apparently a playful side to Hurwitz. There is a wonderful picture of him pretending to conduct an orchestra consisting of two violins, one played by his daughter, Lisi, the other by a young Albert Einstein. This picture was taken around the time that Pólya arrived in Zürich. [1987, 2: 24] Hurwitz had a number of Ph.D. students and was a popular dissertation advisor. It was, however, Hurwitz who said: "A Ph.D. dissertation is a paper of the professor written under aggravating circumstances." [1969, 4]

Adolf Hurwitz (center) with Albert Einstein and Hurwitz's daughter, Lisi

Hurwitz, like Fejér, had a great interest in music—in early life he had had to make a choice, whether to be a mathematician or a pianist. He chose mathematics.

Hurwitz also shared with Fejér a concern for absolute clarity in writing his mathematics. Felix Klein, in his *History of Mathematics in the Nineteenth Century*, called Hurwitz an "aphoristician." As Pólya said, Hurwitz "tended his ideas with loving care, until he arrived at the simplest attainable expression, devoid of superfluous ornament or ballast and transparently clear." Also like Fejér, he "preferred not too large problems which are more amenable to perfect clarity. But his range was much wider than that of Fejér." [1969, 4]

Hurwitz had studied with Hannibal Schubert, the enumerative geometer, in the Gymnasium, and he wrote a joint paper with Schubert when he was a seventeen-year-old student. Later Hurwitz learned "number theory and algebra from Kummer and Kronecker, the Riemannian aspect of complex variables from Felix Klein, the Weierstrassian aspect from Weierstraß himself." [1969, 4] Hurwitz had been appointed a professor extraordinarius in Königsberg (Kaliningrad), where he encountered Hilbert and Minkowski, both of whom became lifelong friends. He had a wide knowledge of mathematics and was called to join the faculty of the ETH in 1892. Upon Hurwitz's death in 1919 Pólya edited his *Mathematische Werke*, in which the appraisal of Hurwitz's work was done by Hilbert himself. Pólya later described Hurwitz's accomplishments as being in the theory of functions, where his dissertation and later papers presented the foundations of the theory of modular functions, independently of elliptic functions. He also discovered a proof of the isoperimetric theorem using Fourier series. In the theory of numbers, he gave two proofs for the fundamental theorem of ideal theory and he developed the number theory of quaternions which was the foundation of his book, *Zahlentheorie der Quaternionen*, published in 1919. It is through a lemma of Hurwitz's that the proof of the four squares theorem using quaternions follows analogously the proof of the two squares theorem using Gaussian integers. In algebra he had developed the generation of invariants by integration.

Pólya wrote: "From the time of my appointment there in 1914 until his [Hurwitz's] death, I was in constant touch with him. We had a special way we worked. I would visit him and we would sit in his study and talk mathematics—seldom anything else—until he finished his cigar. Then we would go for a walk, continuing the mathematical discussion. His health was not too good so when we walked it had to be on level ground, not always easy in the hilly part of Zürich, and if we went up hill, we walked very slowly. I wrote a joint paper with Hurwitz. In fact it is a paper of mine and a paper of his linked in a poetic form of correspondence. My

Adolf Hurwitz

connection with Hurwitz was deeper and my debt to him greater than to any other colleague. I played a large role in editing his collected works."

Pólya reported that Hurwitz kept "a mathematical diary, beautifully, clearly written, well-formulated, where he talked about his ideas and sometimes about his readings, his correspondence and conversations.... Most of it could be printed as it stands. In fact, I tried. I thought of printing parts of the diary, but I never got to it. I quote from it in various papers." (G. Pólya 1978) This practice may have inspired Pólya to do the same. In his personal papers many neatly written volumes of a mathematical diary are found spanning most of his active career. These now reside in the Stanford University Archives.

From Hurwitz Pólya inherited his copy of Riemann's *Gesammelte mathematische Werke*, which falls open to the page where Riemann noted his famous conjecture on the ζ-function!

Soon after Hurwitz's death Pólya tried to move back to Germany by applying for a position at the University of Frankfurt being vacated by Ludwig Bieberbach, who had received the call to Berlin. On January 4, 1921, Pólya wrote Bieberbach a long letter in which he described his recent work in mathematics but also told of the unhappy encounter in Göttingen that had prevented him from going to Frankfurt seven years earlier. He reassured Bieberbach that, contrary to rumor, he was not a Bolshevist, and further, told of the problems he had as a Jew and how this might make his appointment more difficult in Weimar Germany. (G. Pólya 1921)

It is ironic that he should have brought up the fact that he was Jewish with Bieberbach, one of the two most visible mathematicians to embrace the Nazi principles during the Second World War. The other was Wilhelm Blaschke. In later years, Pólya commented that between Bieberbach and Blaschke, he preferred Bieberbach since he at least truly believed in the Nazi cause. Blaschke did not, but cynically used the Nazi party to gain professional advantage and, often, to destroy his colleagues.

"The Joy of the Hunt"

Letters of recommendation to Bieberbach from Hermann Weyl and Erich Hecke survive and each comments on Pólya, along with some other candidates. Weyl wrote:

First of all, there is Pólya. His way of doing mathematics is really completely foreign to me. He is to a lesser degree concerned with

Alle reellen Nullstellen der Polynome

$$1 - 2x + 2x^4 - 2x^9 + 2x^{16} - \cdots + (-1)^{n-1}2x^{(n-1)^2} + (-1)^n x^{n^2}$$

liegen im Intervall $0 < x \leqq 1$.

Keine negativen: $Klar, x | -x$

Keine $> 1 : x | \frac{1}{x}$, nachher $\circ$multiplizieren mit

$\frac{1}{(1-x)^2}$: lauter positive Koeffizienten! Z.B. $n = 4$

```
        16 15        10         5      1 0
         1          -2        +2    -2 1
1/(1-x)  + + + + + + + - - - - - - + + + -
1/(1-x)²  1 2 3 4 5 6 = 6 5 4 3 2 3 4 5 4 4 4 ...
```

NB.

$$\int_0^1 (1 - 2t + 2t^4 - 2t^9 + \cdots) t^{x^2 - 1} dt = \frac{1}{x^2} - \frac{2}{x^2 + 1} + \frac{2}{x^2 + 4} - \frac{2}{x^2 + 9} + \cdots$$

$$= \frac{2\pi}{x(e^{\pi x} - e^{-\pi x})}$$

$$\int_0^1 (1 - 2x + 2x^4 - 2x^9 + 2x^{16} - \cdots + (-1)^{n-1}2x^{(n-1)^2} + (-1)^n x^{n^2}) \, x^{s-1} dx$$

$$= \frac{1}{s} - \frac{2}{s+1} + \frac{2}{s+4} - \frac{2}{s+9} + \cdots + (-1)^{n-1}\frac{2}{s+(n-1)^2} + (-1)^n \frac{1}{s+n^2}$$

$$\frac{1}{(k-1)!} \int_0^1 (1 - 2x + 2x^4 - \cdots \pm 2x^{(n-1)^2} \mp x^{n^2}) \left(\log \frac{1}{x}\right)^{k-1} x^{s-1} dx$$

$$= \frac{1}{s^k} - \frac{2}{(s+1)^k} + \frac{2}{(s+4)^k} - \cdots + \frac{2(-1)^{n-1}}{(s+(n-1)^2)^k} + \frac{(-1)^n}{(s+n^2)^k}$$

Die Anzahl der positiven Wurzeln $(s > 0)$ würde eine untere Grenze ergeben: aber woher wüßten wir das?!

Wesentlicher Grund dieser Formel: vgl. vorne!

Page from Pólya's mathematical log

Hermann Weyl

knowledge but rather with the joy of the hunt. However, I admire his brilliance extraordinarily. His ideas are certainly not of the type that would cast light on the major relationships of knowledge. His papers are rather single, bold advances toward very specific, limited points in an undiscovered land that will remain totally in the dark. But his questions are somewhat unusual. He is full of problems, and is an exceptionally stimulating person in mathematical circles. As an educator he may be somewhat hindered by his anxious desire to temper his investigations to well-defined, precise problems; however, he cares about his students in a way that is best described as a 'sincere fellowship'. As far as applied mathematics is concerned, he is especially strong in probability theory, and has also published in that field. In addition, he is very knowledgeable in applications (physics, statistics, etc.). Overall, he is a very versatile guy (as far as I know, he studied the classics, law, and physics before becoming a mathematician); he is straightforward and doesn't wear blinders. I have the utmost respect for him as a person. He is funny but not a braggart, ebullient (the incident that caused his expulsion from Göttingen was only a dumb outburst of temper);

he is incredibly sincere (this too has hurt him much; he despises apple-polishing types...) and has a strong character. ... He is not particularly close to me; however, he really is one of those people whom I respect the most. (Weyl 1921)

When Weyl says that Pólya is less concerned "with knowledge but rather with the joy of the hunt," he seems to be placing Pólya squarely in Leibniz's camp. Leibniz said, "Nothing is more important than to see the sources of invention which are, in my opinion, more interesting than the inventions themselves."

Hecke recommended Max Dehn for the Frankfurt post, also Radon, Steinitz and Schur. He does not recommend Blumenthal and comments that Pólya is "very sharp but appears to be very artificial." Ironically, he goes on to question whether Pólya could educate students! (Hecke 1921) But, as it turns out, Pólya was found to be unacceptable for an appointment in Weimar Germany, quite possibly due to the fear that he might be a leftist. The position was awarded to Dehn instead.

So Pólya stayed in Zürich, in spite of the heavy teaching load that he complained about to Bieberbach! Some things don't change very much. Professors then, as now, complained of the teaching load. Some years earlier Fekete had written Pólya from Belgrade to complain of having no time for research, due to all the teaching.

In 1914 Hurwitz had been the main attraction in going to Zürich, but there were other interesting people there. Arthur Hirsch (1866–1948) was department head at the ETH and other colleagues included Jérôme Franel (1859–1939), Karl Friedrich Geiser (1843–1934), and Hermann

The Eidgenössische Technische Hochschule

Weyl (1885–1955), who had left Göttingen for Zürich in 1913. Others who would join the department later were Michel Plancherel (1885–1967), Louis Kollros (1878–1959), and Ferdinand Gonseth (1890–1975). Paul Bernays (1888–1977) also left Göttingen to join the faculty at the ETH, but not until 1934. By that time, Weyl had returned to Göttingen (1930) and thence had moved on to Princeton (1933).

Franel is remembered by number theorists for his proof of a necessary and sufficient condition for the Riemann hypothesis involving Farey series, work that prompted Landau to write a couple of papers on the subject. When Pólya knew him, however, he was known primarily as a teacher of mathematics, not a serious research mathematician. Pólya wrote of him fondly. He said of Franel, "He was an especially attractive kind of person and a very good teacher. He gave the introductory lectures on calculus in French for several decades. He had a real interest in mathematics, but he was more interested in French literature. Teaching occupied a good deal of his time but in French literature he had to read everything available. He had no time left to do mathematics. But when he retired he suddenly tackled two of the great problems: the Riemann hypothesis and 'Fermat's Last Theorem.' He asked his good friend Kollros and me one day to listen to his explanation of how he wished to prove the Riemann hypothesis. I listened and tried not to interrupt, but at one point I asked for an explanation. He stopped, was silent for a few minutes, then said, 'Yes, there is the error.'" [1987, 2: 75]

Teaching Evaluations by Shuffling

Geiser had the distinction of being the nephew of the great geometer, Jakob Steiner, who had been at the University in Berlin. In addition, Geiser is remembered for having organized the first internationally recognized International Congress of Mathematicians, in Zürich in 1897. (There had been an earlier congress in 1893 in Chicago as part of the World's Columbian Exposition, but few attended from outside the United States.) By the time Pólya got to Zürich, Geiser was retired. But he still walked several miles every day to the ETH. Pólya remembers that he too told good stories. But Pólya told a story about him.

In German universities there was an anonymous evaluation of the professors: if students liked the class they stomped their feet and if they did not, they shuffled. Geiser was not a popular teacher, so at the end of one of his classes there was an enormous amount of shuffling. When it died down Geiser said very calmly: "May I ask the concerned gentlemen to shuffle with only two feet?" (G. Pólya 1975)

How to Recognize an Ox

Steiner himself had done some very beautiful geometry on the questions of partitions of space, problems that provided Pólya with some material for the first volume of his *Mathematics and Plausible Reasoning.* [1954, 4: 43–58] He also used these questions in his film "Let Us Teach Guessing," which won a film award. Pólya was naturally attracted to the work of Steiner, which he found useful in illustrating his ideas on problem solving, but, in addition, after moving to Stanford in 1942, Pólya collaborated with Szegő in using symmetrization methods of Steiner and Schwarz in order to prove the basic isoperimetric inequality and develop their acclaimed monograph *Isoperimetric Inequalities in Mathematical Physics.* [1951, 2] Steiner had started out very poor and even at the age of sixteen he could not read or write, according to Pólya. He had been a cowherd, but he became a student of Pestalozzi and under his influence developed his intuition. Pólya heard from Geiser this story about Steiner, when he finally became professor of mathematics in Berlin. He was at a party and one of the guests, referred to by Pólya as a "stuffed-shirt," someone with titles, said to Steiner, "Yes, Herr Professor, I hear that you were a cowherd." Steiner responded, "Your excellency, this is so. Since that time, if I see an ox I recognize it right away." (G. Pólya 1975)

With Michel Plancherel, Pólya had a good working and social relationship. They collaborated on two papers, one on mean value theorems [1931, 5] and one on a generalization of the Paley-Wiener theorem. [1937, 2] It is in a way surprising that they got along so well personally. Plancherel was a devout Catholic and a high ranking officer in the Swiss army. Pólya was neither devout nor very sympathetic to the military. But in spite of that, they were good friends.

Pólya also felt close to Paul Bernays, not because Pólya shared Bernays's interest in logic, but because they had a common interest in problem solving. They remained close until Bernays's death. In fact Pólya in later years told of visiting Bernays one day before he died. Bernays was a student of Landau's and started in number theory, but he is remembered today for his work in logic and foundations. Of course, he may be best known for his having seen through to publication Hilbert's ideas on the foundations of mathematics, in the *Grundlagen der Mathematik*. Pólya was in part responsible for having brought Bernays to the ETH from Göttingen.

With Hermann Weyl Pólya shared fewer mathematical interests. As he pointed out, Weyl was interested in great generalizations and Pólya, as we know, preferred to concentrate on concrete cases. Pólya admitted that he and Weyl "never quite understood each other." He did, however, write two papers in Weyl's memory.

An interesting way in which their names are connected is in the famous Pólya-Weyl wager from 1918. It read:

GEORGE PÓLYA and HERMANN WEYL make a bet on the following terms:

Regarding these theorems of contemporary mathematics:

1) Any bounded set of (real) numbers has a precise upper bound,

2) Every infinite set of numbers contains a countable subset.

Weyl makes this prediction:

A   Within twenty years, i.e. by the end of 1937, Pólya himself or a majority of leading mathematicians will admit that the concepts of number, set and countable, to which these theorems refer and which we generally consider basic today, are quite vague; and that inquiries into the validity or falsehood of these theorems are as futile as, let's say, inquiries into the basic truth of Hegel's Philosophy of Nature.

B   Pólya himself or the majority of leading mathematicians will have agreed that the theorems (1) and (2), when interpreted literally in as reasonable terms as possible, are absolutely false (assuming that, either, several interpretations will at that time still be discussed, or that everyone will have agreed on just one interpretation). On the other hand, if by that time it should have been possible to arrive at a clear interpretation of the two theorems, in which at least one of them is true, this will have required a creative effort by which the fundamentals of mathematics will have been given a new and original perspective, and the concepts of number and set will have acquired a new dimension of which as yet we have no notion.

Weyl wins the bet if the forecast comes true; otherwise Pólya wins.

If the two contestants should be unable to agree on who has won the bet, the full professors of mathematics at the Federal Institute of Technology and at the universities of Zürich, Göttingen and Berlin (excepting the contestants) shall be requested to act as a jury, with a decision reached by simple majority. In case of a tie the bet will be considered to remain undecided.

The loser accepts the obligation to publish, at his own expense, in the Proceedings of the German Mathematical Society the terms of the bet and the fact that he lost it.

Zürich, February 9, 1918

(s) H. Weyl   (s) G. Pólya

WITNESSES TO THE VALID MAKING OF THE BET:

Zürich, February 9, 1918

[There were 13 signatures, including T. Carleman, A. Speiser and A. Weinstein.]

The outcome was discussed around the end of 1940 and it was generally conceded that Pólya had won, though Pólya reported that Weyl thought that he was 49% right, Pólya 51%. Weyl did ask, however, that he not be held to the consequences of the wager and Pólya graciously agreed. (G. Pólya 1978)

Monkeys and Mathematicians

Pólya would have had some of the same problems with Weyl's work that he had in appreciating the work of Emmy Noether, who was also a proponent of generalization. He reported that he once interrupted her with the remark: "Now, look here, a mathematician who can only generalize is like a monkey who can only climb up a tree." She was hurt by the remark and Pólya regretted having made it. But he decided later that she should have responded, "And a mathematician who can only specialize is like a monkey who can only climb down a tree." Pólya said, "A real mathematician must be able to generalize and to specialize. A particular mathematical fact behind which there is no perspective of generalization is uninteresting.

On the other hand, the world is anxious to admire that apex and culmination of modern mathematics: a theorem so perfectly general that no particular application of it is possible." [1973, 3]

Two other colleagues at the ETH with whom Pólya had interesting discussions were the physicists Wolfgang Pauli and Fritz Zwicky. He did admit that he could not discuss physics very much with Pauli because he was not up to it—he did not have the necessary knowledge. But he did discuss mathematics with him. Pointing out that Pauli did not hide his opinions and they could be very sharp, he said, "Once I heard a discussion with Weyl and he [Pauli] said to Weyl: 'Your writings are covered with makeup as thick as a finger. Before one reads them, one has to rub the makeup away' and Pauli showed very visibly how to rub it away. And Weyl sat and smiled ...." He reported on another encounter of Pauli's, this time with Zwicky, who later joined the faculty at the California Institute of Technology. "Pauli said to Zwicky: 'Dear Zwicky, wouldn't you once propose a theory which can be tested experimentally?' And Zwicky was so hurt by this remark which had some point, that all that he could answer was, 'You are a liar.'" (G. Pólya 1978)

Prior to Pólya's arrival in Zürich in the spring of 1914, he had published fifteen papers, not a bad start for someone who had earned his Ph.D. two years earlier. Some of these, however, are near duplicates in that some were published in Hungarian but then, subsequently, German versions of them appeared. One was his dissertation which concerned some definite integrals arising in geometric probability. He did not appear to pursue questions of that sort immediately after his degree but instead moved into some questions of analysis. His first major paper was a long piece coauthored with Mihály Fekete on a problem of Laguerre. [1912, 4] The problem concerns the number of sign variations in a sequence of coefficients of a series formed in a particular way from a real polynomial. The article consists of a series of letters between Pólya and Fekete between December 1911 and February 1912. Both were in Budapest at the time. Pólya was to return to this problem in 1928, and others, including I. J. Schoenberg, were inspired to look at aspects of this problem as well.

This was the first of a long series of papers running over many years concerning zeros of polynomials and "sign rules for zeros." In [1913, 3] Pólya gave the first correct proof of Laguerre's theorem on Laplace transforms. He was also looking at Graeffe's root-squaring method [1913, 6], an idea he was to return to as late as 1968.

But in the midst of doing all this work on the location of zeros, he took time out to publish a paper on a Peano curve. [1913, 5] Pólya's curve is a variant of the famous space filling curve of Sierpiński. What makes Pólya's curve interesting is that, remarkably, the curve passes at most three times through any point in the interior of the scalene triangle over which it is defined, a possibility hinted at earlier by Hilbert. Later, in 1933, Witold Hurewicz showed that triple points must exist in a Peano curve, but Pólya's curve is remarkable in that the curve does not have points that the curve passes through more than three times. [1987, 2, vol. 3: 485] Conditions on the triangle under which this curve is nowhere differentiable were investigated by Peter Lax in 1973 and Richard Bumby in 1975. (Sagan, 1994: 62–68) For details on the construction of the curve, see Appendix IV.

In 1915 Pólya published what was to be one of his most admired results in analysis, the so-called 2^z theorem, which says that 2^z is the "smallest" (in a well-defined sense) transcendental integral function with integral values at the nonnegative integers. [1915, 2] This was to lead to a number of related results, by Atle Selberg, Ludwig Bieberbach, and by Pólya himself. (See [1974, 3, vol. 1: 771–772] and Appendix II.) Halsey Royden in 1977 was to describe this as one of his two favorite Pólya theorems—the other being the Pólya-Carlson theorem. (Royden 1977) This latter theorem had been conjectured in a paper Pólya published in 1916 [1916, 3], but it was not confirmed until Carlson offered a proof in 1921. (Carlson 1921) This theorem states that a power series with integral coefficients and radius of convergence 1 represents either a rational function or a function with the unit circle as the natural boundary. This too had consequences that were explored further by Pólya and others. A p-adic version appears in number theory in the work of Bernard Dwork.

Royden wrote of Pólya's work on entire functions: "Pólya's most profound and difficult work is in the theory of functions of a complex variable. He was one of the pioneers, along with Picard, Hadamard, and Julia, of the modern theory of entire functions. It is an indication of the level of Pólya's contribution that the language of the subject contains such phrases as 'Pólya peaks', 'the Pólya representation', 'the Pólya gap theorem', 'the Pólya-Carlson theorem', 'Pólya's 2^z theorem', etc. Some of Pólya's most interesting work in this area concerns the zeros of entire functions. One paper of Pólya's in 1926 came close to proving the Riemann hypothesis. Although it failed to do so, it led to further developments, including some in statistical mechanics." (Royden 1989)

It was in those early years in Zürich that Pólya took up the question of the location of zeros of Fourier transforms and the Riemann ξ-function, knowing that a good result in this area could lead to a proof of the Riemann hypothesis. [1918, 4] By the late 1920's people felt that he was moving close to a proof of significant results on the ξ-function. [1926, 8] Although these results on the ξ-function did not lead to a proof of the Riemann hypothesis, they did have unexpected influence on C. N. Yang and T. D. Lee in developing their theories in statistical mechanics. [1974, 3, vol. 2: 424–26]

In 1917–18 Pólya wrote two papers that were to stimulate a good deal of further work. They concerned the problem of determining the size of the largest prime factor of certain polynomials. [1917, 3] [1918, 2] Work on this problem extends to the latter part of this century, with contributions by Kurt Mahler, S. Chowla, Alan Baker and John Coates. Also at about this time, stimulated by work of Edmund Landau and Issai Schur, Pólya did some interesting work on quadratic residues and nonresidues. [1918, 3]

As was the case for many first-class mathematicians of his day, though, the Riemann hypothesis held enormous appeal. With Pólya it was a mild obsession. Proving it would have been for the prover, a stunning achievement. The Riemann hypothesis was always there and Pólya was always pondering over it. Up until a few days before his death, when he was in and out of consciousness and generally confused about his surroundings, he was convinced that he had come close to proving the Riemann hypothesis in a "recent" paper and he wanted to see it. Of course, we all looked for it but no such paper existed.

Zürich and Marriage

When Pólya came to Zürich he moved into a room at the Hotel (*Kurhaus*) Zürichberg which is on a mountain above Zürich, in a wooded setting, close to the well-known Fluntern cemetery. (This cemetery was in later years a pilgrimage spot for admirers of James Joyce, who was buried there until his bones were eventually moved to Ireland.) The Hotel was and is a large establishment, founded as a temperance hotel (*Alkoholfrei*). It provided comfortable but rather modest accommodations, fairly convenient to the ETH. It was here that Pólya met his future wife, Stella Vera Weber.

Stella's father, Robert Weber, had been a professor of physics at the University of Neuchâtel, but he had suffered a stroke quite early in life,

Stella Pólya (right) with mother and two sisters

at the age of 53, and was no longer able to function at the University. The circumstances of the family had been comfortable in Neuchâtel but became difficult with the father's debilitating illness. Weber decided to spend his time traveling, leaving his wife and their four children back in their apartment in Neuchâtel, where Stella's mother was forced to take in lodgers. He spent at least part of each year at the Hotel Zürichberg, in his home canton. It was there that Stella would visit her father from time to time and where she met the young George Pólya. She observed that he would sit with her father for hours playing checkers, a game that she thought unbelievably tedious. This exhibition of kindness on Pólya's part impressed her and that was the beginning of a relationship that would lead to their marriage in 1918.

They announced their engagement in August of 1916, but the engagement was to last for two full years since, somewhat earlier, Stella's brother and two sisters had died of tuberculosis, and the family physician decided that Stella was also susceptible to the disease. So she was sent off to Davos for two years to stay in a sanatorium and breathe the fresh mountain air. She described this as an incredibly boring period in her life, since she had little to do but read and look at the mountains.

Stella Pólya's mother, George Pólya, Anna Deutsch Pólya, Ilona Pólya

The town was full of sanatoria for tuberculosis patients. In earlier times it had been the temporary home of a number of consumptive writers, including John Addington Symonds and Robert Louis Stevenson. Pólya would visit her there on holidays; otherwise they stayed in touch only through correspondence. Eventually she was allowed to leave and they were married on August 1, 1918. Perhaps significantly, August 1 is the Swiss national holiday.

Random Walk — Dividend of Not Driving

After their marriage they continued to live at the Kurhaus Zürichberg, and it was probably a fortunate choice, because Pólya found the wooded surroundings appealing. For all of his life he very much enjoyed walking. In fact, he never learned to drive, so he was always dependent on walking, being driven somewhere by others or using public transportation. (He put it this way: "I hold a pedestrian license.") There was a particular wooded area near the hotel where he enjoyed taking extended walks while thinking about mathematics. He would carry pencil and paper so he could jot down ideas as they came to him. Some students also lived at the Kurhaus and Pólya got to know some of them. One day while out on his walk he encountered one of these students strolling with his fiancée. Somewhat later their paths crossed again and even later he encountered them once again. He was embarrassed and worried that the couple would conclude that he was somehow arranging these encounters. This caused him to wonder how likely it was that walking randomly through paths in the woods, one would encounter others similarly engaged. [1970, 2] This led to one of his most famous discoveries, his 1921 paper on random walk, a phrase used for the first time by Pólya. [1921, 7] The concept is critical in the development of the investigations of Brownian motion.

In order to put the problem in manageable form, Pólya considered an infinite plane with a grid of equally spaced lines meeting at right angles. This would be the map of an idealized and rather boring town where all the blocks are the same size, the streets meet at right angles, and the town extends indefinitely in all directions. (For someone in Switzerland this simplification would probably have been seen as a stretch of the imagination, since normally Swiss towns are nowhere near being as regular as this. The *Guide Michelin* for Switzerland once referred to the exception, La Chaux-de-Fonds, the Swiss home of Omega watches and a city that does exhibit such regularity, as being constructed *à l'américaine*, a damning description indeed!) But let's imagine such an "American" town where a man stands on a certain street corner and has the possibility of moving in any one of four directions, north, south, east, or west. Suppose that the probability is the same of moving in any of the four directions, that is, in each direction the probability is 1/4. Suppose that after walking one block the person faces exactly the same decision, whether to walk another block north, south, east, or west, and again the probability of moving in any of the four directions is 1/4, and so on. This is a simple random walk in two dimensions. (This has been referred to as the Pólya drunkard problem since the man walking from block to block in this manner could be viewed as a drunkard trying to find his way home.)

A one-dimensional random walk would involve starting at a point on a line and moving backwards or forwards a unit length, with probability 1/2 of moving in either of the two directions. A possible interpretation of this latter problem would be the fate of a gambler who is betting heads or tails on repeated tosses of a fair coin. After a certain number of tosses, corresponding to moves of unit distance in one direction (heads) or the other (tails) along the line, the final position of the moving point would correspond to the total gain or loss of the gambler. This is the "gambler's ruin" problem of Laplace. Stated more precisely, we can say that the probability of returning to the original point (and, hence, infinitely often) is 1 in one and two dimensions. Curiously enough, not only is the probability not 1 in three dimensions, it drops down to approximately 0.35. Furthermore, if one has two particles moving in a random walk, as in the motivating case of Pólya and the young couple walking in the woods, in two dimensions, the probability of meeting infinitely often is 1, but in three dimensions there is a positive probability they will never meet (Feller 1968, 360).

Pólya remarked that in two dimensions, all roads really do lead to Rome!

Naturally one can also generalize this problem to a lattice in d-dimensions with the mutually perpendicular lines connecting points with integer coordinates. At each vertex, that is, at each point of intersection of the lines, there would be $2d$ edges from which to choose, with a probability of $1/2d$ of moving in any one of the directions, that is, along any of the edges. In Pólya's paper on random walk (originally given the French name, "*promenade au hasard*"), he proved an amazing theorem. [1921, 7] In a one- or two-dimensional lattice, the moving point *must* return to its starting position given sufficient time. But this need not happen in any higher dimension.

~ 5 ~

Collaboration with Szegő

Aufgaben und Lehrsätze

In the first few years after Pólya's arrival in Zürich, he produced what is really an amazing amount of good work. The quantity and range are both impressive. Seven papers appeared in 1914, five in 1915, six in 1916, nine in 1917, nine in 1918, and thirteen in 1919. In 1918, for example, he published papers in number theory, combinatorics, series, and even on voting systems. The next year there were papers on analysis and number theory and, again, on voting systems, but in addition there were other papers on probability, and even papers on astronomy and pedagogy. During this time he was starting to obtain deep results in the area of entire functions.

A paper of 1919 was his first venture into the field of problem solving and pedagogy. It was probably Pólya's interest in making mathematics completely clear—the influence of Fejér and Hurwitz—that prompted his initial interests in education. But already in 1917 he was considering questions of teaching, since on November 22 of that year he spoke on mathematical discovery at the town hall in Zürich. The ideas in this presentation provided the content for his first paper on problem solving. [1919, 6] Although the ideas in this paper and a few other early papers were supplanted by later, more comprehensive and more profound work in the subject, nevertheless it is interesting to note that a diagram appearing on the end paper of the second volume of his *Mathematical Discovery* [1962, 1] is clearly an outgrowth of a diagram in his 1919 paper.

A far more influential work of this period was, of course, the problem collection assembled by Pólya and Szegő, the *Aufgaben und Lehrsätze aus der Analysis* (*Theorems and Problems from Analysis*). By 1982, when Pólya was asked to write about his collaboration with Gábor Szegő, he was no longer able to remember whether he had met Szegő in 1912 or in 1913. At any rate, the meeting took place in Budapest after Pólya had received his doctorate and was on the postdoctoral appointment in Göttingen. Szegő was a student at the University of Budapest at that time, though he received his Ph.D. in 1918 from the University of Vienna, after having done military service during World War I. His dissertation was written when Szegő was 20 and fighting in the trenches. After a few temporary positions Szegő was appointed a *Privatdozent* in Berlin in 1921. Then, after one year, he moved on to the University of Königsberg, where he became Professor of Mathematics in 1926, leaving in 1934 only because the Nazis made it impossible for him to continue to work in Germany.

Szegő was born on January 20, 1895, in Kunhegyes, Hungary. So he was roughly seven years younger than Pólya. In 1912 he had won the Eötvös Competition and his comments on this appear in the collection of Eötvös problems. (Kürschák 1963) Szegő's first publication in an international journal was a solution of a problem posed by Pólya in the *Archiv der Mathematik und Physik* in 1913 (Zu 424, *Arch. Math. Phys.* 21 (1913), 291–292).

In 1914 Pólya published in *L'Intermédiaire des Mathématiciens* a conjecture on the solution of a problem on determinants of Toeplitz matrices, a problem that had been looked at by Toeplitz, and by Carathéodory and Fejér as well. Szegő proved this conjecture in his first research paper, "Ein Grenzwertsatz über die Toeplitzschen Determinanten einer reelen positiven Funktion," published in the *Mathematische Annalen* in 1915. This was the first of a whole series of Szegő's papers in this general area. The other field in which Szegő produced important research was the theory of orthogonal polynomials on which he wrote an AMS Colloquium volume in 1939 (revised in 1958).

Pólya himself described his conjecture on Toeplitz determinants and Szegő's solution thus:

> Our cooperation started from a conjecture which I found. It was about a determinant considered by Toeplitz and others, formed with the Fourier-coefficients of a function $f(x)$. I had no proof, but

I published the conjecture and the young Szegő found the proof.…
We have seen here a good example of the fruitful cooperation between two mathematicians. Mathematical theorems often, perhaps
in most cases, are found in two steps: first the guess is found; then
minutes, or hours, or days, or weeks, or months, perhaps even several years later, the proof is found. Now the two steps can be done
by different mathematicians, as we have seen. (Szegő, 1982, 11)

"No Mere Collection of Problems"

Pólya clearly respected Szegő's power as a mathematician and realized
that Szegő was familiar with some areas of mathematics that Pólya was
not, so he proposed a collaboration on a book of problems. In 1923 they
signed a contract with Springer-Verlag for what was to be their great
classic work, the *Aufgaben und Lehrsätze aus der Analysis*. Unlike the case
with many collections of problems, here Pólya and Szegő organized the
problems so that whole areas of mathematics could be developed through
the careful organization and ordering of the problems themselves.

Szegő and Pólya in Berlin at the time of signing with Springer for the Aufgaben
und Lehrsätze

Though the book, as the title indicates, deals with analysis, it also contains a good deal of combinatorics, geometry and number theory, as well as some applications to physics. Generations of graduate students have learned the fundamentals of analysis by systematically going through the problems in the *Aufgaben und Lehrsätze*.

What Pólya and Szegő set out to do is probably best described in their own words, in their Preface to the first German edition:

> The chief aim of this book, which we trust is not unrealistic, is to accustom advanced students of mathematics, through systematically arranged problems in some important fields of analysis, to the ways and means of independent thought and research. It is intended to serve the need for individual active study on the part of both the student and the teacher. The book may be used by the student to extend his own reading or lecture material, or he may work quite independently through selected portions of the book in detail. The instructor may use it as an aid in organizing tutorials or seminars.
>
> This book is no mere collection of problems. Its most important feature is the systematic arrangement of the material which aims to stimulate the reader to independent work and to suggest to him useful lines of thought. We have devoted more time, care and detailed effort to devising the most effective presentation of the material than might be apparent to the uninitiated at first glance.
>
> The imparting of factual knowledge is for us a secondary consideration. Above all we aim to promote in the reader a correct attitude, a certain discipline of thought, which would appear to be of even more essential importance in mathematics than in other scientific disciplines.

"Try to Understand Everything"

Later they go on to say:

> One should try to understand everything: isolated facts by collating them with related fact, the newly discovered through its connection with the already assimilated, the unfamiliar by analogy with the accustomed, special results through generalization, general results by means of suitable specialization, complex situations by dis-

secting them into their constituent parts, and details by comprehending them within a total picture.   [1972 [1925, 3], 1: vi–vii]

This paragraph is prescient of Pólya's later books on problem solving, in particular his *Mathematics and Plausible Reasoning,* where there is a good deal of emphasis on generalization, specialization, analogy, and such, and the later *Mathematical Discovery,* where the same principles are emphasized, albeit for more elementary problems. It is clear that as early as 1925, Pólya was well on his way to describing the principles of problem solving that he was so eloquently to espouse later in his career.

Further along in the Preface the authors give a rather homely description of generalization and specialization, again a theme that would arise later:

> One must not forget that there are two kinds of generalization, one facile and one valuable. One is generalization by dilution, the other is generalization by concentration. Dilution means boiling the meat in a large quantity of water into a thin soup; concentration means condensing a large amount of nutritive material into an essence. The unification of concepts which in the usual view appear to lie far removed from each other is concentration. Thus, for example, group theory has concentrated ideas which formerly were found scattered in algebra, number theory, geometry and analysis and which appeared to be very different. Examples of generalization by dilution would be still easier to quote, but this would be at the risk of offending sensibilities. [1972 [1925, 3], 1: viii]

The effect of these two volumes was so profound that after publication of the books in 1925, the names Pólya and Szegő were henceforth connected in the minds of mathematicians. The combination is even alluded to by Katharine O'Brien in a poem that she entitled "Aftermath." In this she worries about what will happen to her mathematical books and papers when she leaves this world and moves on to the next. One of the stanzas reads:

> Now Hardy's a treasure and Banach a pleasure
> and the Knopps a delight for the mind.
> There is Pólya *und* Szegő—well, I go where *they* go—
> couldn't bear it to leave them behind.   (O'Brien 1967, 36)

M. M. Schiffer remarked that Pólya and Szegő go together like Gilbert and Sullivan or Hardy and Littlewood!

"It Is Not Forbidden to Learn ... from an Examination"

The *Aufgaben und Lehrsätze* is a very complex set of books with an amazingly subtle structure, where problems are related within sections, but also there are relationships connecting the sections. The list of reviewers of the 1925 edition was itself impressive: Frigyes Riesz, Konrad Knopp and J. D. Tamarkin. For a full analysis of the contents and impact of these volumes, see the review by Paul C. Rosenbloom. (Szegő 1982, 1: 23–32) This is a review of the English edition, somewhat expanded, that appeared in 1972 and 1976 (*Problems and Theorems in Analysis I and II*, Springer-Verlag). In another essay in those volumes, Professor Rosenbloom describes experiences as a graduate student at Stanford in the 1940s. When Rosenbloom was looking for a dissertation topic, Szegő suggested that he wait for Pólya to arrive—which he was to do in the spring of 1942. At that point, according to Rosenbloom, Pólya suggested that he look "to prove the results on sections of power series for entire functions which Carlson had announced in 1924, whose proofs had never been published.... At one point, when I was floundering, Pólya suggested that I try to apply problems 107–112 in the first volume of Pólya-Szegő, and these turned out to be crucial. I was soon able to handle entire functions of infinite order and the radial distribution of the zeros of sections of entire functions of finite order.... When I told Pólya my results, he invited me to his home. The whole evening he pestered me with questions about why my proof worked, whether my solution was a special trick or an instance of some systematic method. I learned more from that one evening than from any other single experience in my career." Rosenbloom goes on to say that during his orals at Stanford, "Pólya began asking me to give the definition of Gaussian curvature in terms of the area of a spherical map by the normals. I protested that I had never studied it, but he insisted that I try to work it out at the blackboard. He said, 'It is not forbidden to learn something from an examination!'" (Szegő 1982, 1: 12–13)

I had a similar experience with Pólya much, much later in his career, when I asked him a question about binomial coefficients. As it turns out, the day I saw him he was not at his best. A couple of days earlier he had fallen in a trench at a construction site at Stanford on the way to his office and had cracked several ribs. So he had been sitting up all night.

He had had very little sleep and was in a lot of pain. But when he heard the problem he thought for a while, said that he had never seen it before, but cited a problem in Pólya-Szegő that would do the trick. And it did. On another occasion, when considering the problem suggested by Kürschák's tile (which demonstrates by equidecomposability that the regular dodecagon inscribed in a unit circle has area equal to 3) (Alexanderson and Seydel, 1978), on the rational nature of areas of inscribed and circumscribed regular polygons in or about a unit circle, again Pólya was able to cite a set of problems in Pólya-Szegő that would resolve the problem completely—the problem this time involved questions concerning rational values of trigonometric functions.

Most Useful Problem Book

A person could probably spend a good part of a lifetime investigating the consequences of the problems in the *Aufgaben und Lehrsätze*. Richard Askey and Paul Nevai have stated in their work on Szegő that "there is general consensus among mathematicians that the two-volume *Pólya-Szegő* is the best written and most useful problem book in the history of mathematics." (Askey and Nevai, 1996)

Hidden away in the two volumes are little gems that have led to series of interesting investigations. For example, in Volume 2, Chapter 5, Problem 239, one finds the following question: "How thick must the trunks of the trees in a regularly spaced circular forest grow if they are to block completely the view from the center?" This question had been addressed earlier by Pólya in a more extensive paper on number theory and probability. [1918, 5] Now known as Pólya's Orchard Problem, it has prompted various investigations. (Honsberger 1973, 43–53; Allen 1986) Honsberger's approach to the problem uses Minkowski's Convex Body Theorem; Pólya's is based on a method of Andreas Speiser's.

Here's another example. Halsey Royden said that from graduate school he knew two proofs that there are infinitely many primes, the first by Euclid, the second by Pólya. (Royden 1977) The latter proof is an answer to Problem 94 in Part VIII of Volume 2, Chapter 2, of the *Aufgaben und Lehrsätze*. Hardy and Wright included this proof in their *Theory of Numbers* and it was recently cited again by Ribenboim in an article on prime number records for which he appropriately won the Pólya Award for expository writing given by the Mathematical Association of America.

(Ribenboim 1994) The proof involves showing by a simple argument that any two Fermat numbers are relatively prime. The nth Fermat number equals $2^{2^n} + 1$. If $m|F_n$ and $m|F_{n+k}$, $k > 0$, since

$$F_{n+k} - 2 = \left(\left(2^{2^n}\right)^{2^k -1} - \left(2^{2^n}\right)^{2^k -2} + \cdots - 1 \right) F_n$$

we conclude that $m|2$. But Fermat numbers are all odd, so $m = 1$. Hence, since there are infinitely many Fermat numbers, no two of which share a prime factor, there must be infinitely many primes.

Saving the New Generation

Even at the time of publication these volumes were recognized as likely classics. Donald E. Knuth, on the occasion of the 50th anniversary celebration of the *Aufgaben und Lehrsätze*, summarized the initial reviews. Frigyes Riesz found them "the first really interesting texts for advanced students of mathematics. He said that the early years of mathematical training were full of stimulating problems for building up intuitive power, but then the advanced books were deathly dull; so, he said, Pólya and Szegő have saved the new generation from this disillusionment and boredom, by their masterful choice and arrangement of the material." Konrad Knopp said "that the books' value couldn't be too highly estimated; and he said that Pólya and Szegő should please write a third volume about Bessel functions, Tauberian theorems, divergent series, and a few other things." J. D. Tamarkin said "there are but few books which could be compared with this one as to the richness and charm of material." (Knuth 1974)

During the period when Pólya and Szegő were working so intensively on these volumes, and Pólya was, in addition, writing research papers in a variety of fields, Stella Pólya worried about his health and thought that under the enormous stress involved, he could suffer a breakdown. Her concerns were probably justified. Pólya's 1921 letter to Bieberbach asking for support in his application to succeed Bieberbach in Frankfurt reveals, in tone, an agitated state of mind, an air almost of desperation. It is the letter of a man who is married and with a sense of obligation to wife and family, but who is underpaid, asked to teach too much, and forces himself to work constantly on his research and writing.

In this letter Pólya informs Bieberbach that he understands that, because of various factors, he has little chance of an appointment in

*Gábor Szegő and his son
Peter in Königsberg*

Germany. He talks about the incident on the train, described earlier, his failure to report to serve in the armed forces, and he addresses the fact that his family is Jewish. "My father was indeed a Jew. He accepted baptism a few years before I was born, and therefore I was, as is evident from all my papers, always officially a Roman Catholic. (It won't be necessary to explain my unofficial views.)" He goes on to describe his situation as a *Privatdozent*: "In Switzerland I have substandard prospects ... I have a vague hope for (a position in) Holland, England, and North or South America. If the system changes in Hungary, I would also assume a subordinate position there.... It impresses me very little, when people who are considerably inferior to me in accomplishments and education are given preference to me.... I consider that my salary is not only less than that of a trolley conductor, but in addition this income is completely uncertain, and from this income I am supposed to support my wife.... for the past seven years I have had to repeat up to 20 hours per week material for the least advanced students." He apologizes to Bieberbach for his outburst.

But the outburst clearly derives from frustration and overwork. The intensity of his application to his work is alluded to by Hardy in a letter quoted in the next chapter.

It is somewhat surprising that Pólya should be so eager to leave Zürich in 1921. While remaining a *Privatdozent* he would certainly have had financial concerns, but in 1920 he had been promoted to Titularprofessor and he was named Professor Ordinarius in 1928. It was a promotion supported by Hardy in the letter referred to above.

In May of 1919 Pólya had actually received an appointment to the Faculty of Humanities in Budapest, in the division of the University devoted to the teaching of Gymnasium teachers. The appointment had come from György Lukács, the Hungarian writer then in the cabinet of the Soviet-style government of Béla Kun, which had overthrown a Kerensky-style government of Count Károlyi. The list of new appointments to this faculty that year was impressive, including Karl Mannheim, the sociologist and historian, Arnold Hauser, art historian and French linguist, and Marcel Benedek, sociologist. Mathematicians appointed at that time were Fekete and von Kármán. However difficult the situation was in Zürich, the decision not to move to Budapest was almost certainly the correct one, since the government of Kun's was short-lived, succeeded by the right-leaning government of Nicholas Horthy de Nagybánya. Pólya would not have been comfortable under Horthy's government, which later agreed to Hungary's entering the Second World War on the side of Germany. (Farkas, 1967)

The fact that the Pólya and Szegő collaboration is so well-known, but that they collaborated on little else in the early years, is evidence of the importance of this famous two-volume problem book. Their names may be forever juxtaposed, but they did not collaborate on any research papers until 1931, and then only on two. There was little mathematical contact till they resumed their collaboration in the 1950s, when both were on the faculty at Stanford University. Prior to that time, however, a collaboration would have been somewhat difficult since they were geographically separated. This did not, on the other hand, seem to get in the way of their collaboration on the *Aufgaben und Lehrsätze* when one was in Königsberg, the other in Zürich. Pólya in commenting on the work said that the book " ... is my best work and also the best work of Gábor Szegő." (Szegő 1982, 1: 11)

6

Oxford and Cambridge

G. H. Hardy

In 1924, one year after signing the contract with Springer for the *Aufgaben und Lehrsätze,* and one year before its publication in 1925, surely a busy year for Pólya, he applied for a special fellowship from the Rockefeller Institute to work in England. G. H. Hardy had been approached by the Rockefeller staff and had recommended Pólya for one of the first fellowships. On July 3, 1924, Hardy wrote from New College, Oxford,

> Dr. G. Pólya has been nominated by the International Education Board of the Rockefeller Institute to spend a year studying mathematics in England at Oxford and Cambridge. His expenses in England will be defrayed by the Institute, and he will occupy his time in scientific research (nominally under my direction). I trust that all possible facilities will be given to Dr. Pólya to enable him to accept this invitation.
>
> (signed) G. H. Hardy, Savilian Professor of Geometry in the University of Oxford.

Pólya was greatly indebted to the International Education Board and was anxious to enjoy the contact with Hardy and Littlewood, but he was also eager to have a year he could devote exclusively to research. Further, he remarked in a letter to Hardy that it would allow him an opportunity to give presentations to English audiences, "which could have importance for my future." Indeed, his ability to handle English was important to his future, as we shall see later. Hardy viewed Pólya's stay in England as a success and so informed Augustus Trowbridge, director of the International Education Board in Paris in a letter dated March 13, 1925. (Rider 1984)

Hardy had earlier been at Trinity College, Cambridge, but had moved to Oxford in 1920 when he was elected to the Savilian Chair. He was not to return to Cambridge until 1931 when, upon the death of E. W. Hobson, he became again a Fellow of Trinity and was elected to the Sadleirian Chair of Pure Mathematics there. This moving back and forth between Cambridge and Oxford led to a question about his loyalty to the two universities. First one must know that Hardy loved cricket but despised rowing. Pólya remarked that when someone asked Hardy, "For which university are you in sports?", Hardy replied, "It depends. In cricket I am for Cambridge, in rowing I am for Oxford." [1969, 4]

In 1912 Hardy had begun his long and productive collaboration with J. E. Littlewood, and in 1913 he received the now famous letter from Srinavasa Ramanujan which resulted in Ramanujan's coming to England in 1914, where he remained until 1919, the year he returned to India. He died there in 1920. It is interesting to note that Ramanujan and Pólya were born in the same year, 1887. By the time Pólya came to Cambridge, though, Ramanujan had been dead for four years.

Pólya with Hardy at Oxford

Some Very Special Cricket Teams

Hardy was extraordinarily fond of cricket and his highest compliment about anything was to say, "It is in the Hobbs class." (Titchmarsh, 1949) This referred to Sir John Berry Hobbs (1882–1963), the famous English cricketer. Hardy was so fond of cricket and of making up lists that one of his pastimes was to make up lists of members of cricket teams. These were highly fanciful. Among Pólya's papers were five such teams, written out in Hardy's distinctive hand (the various references to God are, we must assume, to God the Father, God the Son, and God the Holy Ghost):

First team: Hobbs
                   Archimedes
                   Shakespeare
                   M. Angelo
                   Napoleon (Capt)
                   H. Ford
                   Plato
                   Beethoven
                   Johnson (Jack)
                   Christ (J)
                   Cleopatra

Jack Johnson was an African-American boxer. M. Angelo and H. Ford are, presumably, Michelangelo and Henry Ford.

Second team: D. [sic] Spinoza
                   A. Einstein
                   David
                   B. Disraeli (Capt)
                   God (F)
                   H. Stinnes
                   P. G. H. Fender
                   E. Lasker
                   Paul
                   God (S)
                   God (H.G.)

Hugo Stinnes was a German industrialist, P. G. H. Fender a cricketer and Emmanuel Lasker a chess player and mathematician. From this list

A couple of cricket teams

and the preceding we conclude that Hardy clearly recognized the leadership capabilities of Napoleon and Disraeli.

Third team: P. G. H. Fender (Capt)
D. Spinoza
A. Einstein
A. de Rothschild
Moses
David
H. Heine
B. Disraeli
God (F)
God (S)
God (H.G.).

This appears to have been an all Jewish team. The identity of everyone is quite clear except, perhaps, for the poet Heinrich Heine.

The fourth and fifth teams were made up for Hardy and Pólya, all the members of one team having their names begin with "Ha" and all the members of the other team made up of names beginning with "Po".

Hayward (T)	Poincaré (H)
Hannibal	Porsena
Haydn	Pontius Pilate
Hakon	Poe
Hamilton (Sir WR)	Poisson
Hardy (T.)	Potiphar (Mrs.)
Hafiz	Poincaré (R)
Hapsburg (R von)	Poushkin
Harmodius	Pond
Hamlet	Poinsot
Hadamard	Polycrates

T. Hayward was a cricketer. Porsena is almost certainly Lars Porsena of Clusium (See Macaulay's *Lays of Ancient Rome*.) Some of these take a bit of interpretation. The spelling is odd on Hakon, but this is presumably the king of Norway. Hafiz is the fourteenth century Persian poet. The Hardy is Thomas Hardy, to distinguish this from G. H. Hardy himself. The story of the double suicide of Rudolf von Hapsburg and the Baroness Maria Vetsera at Mayerling were still a bit fresher in people's minds then than they are today. To choose an Athenian assassin for one's team would take some courage, but there's Harmodius. Non-mathematicians are more likely to know Raymond Poincaré, the president of France, than his cousin, Jules Henri, the mathematician. Here we have both. This is clearly Potiphar's wife, from the Joseph story in Genesis. Pushkin's name is possibly given a French transliteration to get the "Po" needed. Poisson and Poinsot were mathematicians, of course, and Polycrates is remembered not for his mathematics but for what he did to a mathematician: he drove Pythagoras from Samos.

Cricket was not the only game in which Hardy was intensely interested, though nothing mattered quite as much as cricket. C. P. Snow was to write about this interest in cricket in an article entitled "A mathematician and cricket" that appeared in the *Saturday Book, 8th Year*. Also in these papers was a four-page handwritten account by Hardy of the rules of tennis as played in the court of the seventeenth century French kings.

Pólya's visit to England in 1924–25 led to a long friendship between Pólya and Hardy that lasted until Hardy's death in 1947. Pólya once remarked that Hardy was the mathematician who impressed him most personally. He said, "Hardy had quite a special personal charm. I cannot describe it, what the charm was, but everyone was attracted by him, men and women, mathematicians and nonmathematicians, intellectuals and very simple people."

Tripos

At the time of Pólya's visit to England, Hardy and Littlewood were trying to reform the famous mathematical Tripos examination at Cambridge, which they felt was made up of problems that were difficult but no longer relevant to serious, modern mathematics. In order to prove their point Hardy convinced Pólya to take the test. He assumed that Pólya would do badly and this would demonstrate that the mathematics on the Tripos was irrelevant, since a leading young mathematician from the Continent had had trouble with these artificial, contrived problems. The demonstration failed since, had Pólya been officially entered, he would have been named Senior Wrangler! (Royden 1989)

Hardy was clearly fond of Pólya and, in particular, Hardy got along very well with Stella Pólya. Hardy's relations with women were not always easy. He could be shy around women. But he seemed to have a particular rapport with Mrs. Pólya and they were often partners at bridge. There is a curious reference to Stella Pólya in a letter of Hardy's to Pólya, written from Trinity College. There is no date on the letter, but since it's from Trinity it must date from after 1931. There are some references to poetry and a question concerning the American poet, Ella Wheeler

Stella Pólya

Wilcox. It is not clear why Hardy would have thought that Mrs. Pólya would know these lines of poetry since she was Swiss and not American, not even native English-speaking. Mrs. Pólya thought that it possibly was a comment meant for Mrs. G. D. Birkhoff. Here is the letter.

Ranking Poets

Trinity College
Cambridge
30 Nov.

Dear Pólya

This is merely to acknowledge the MS.

I am giving it to White. It is possible he may prefer to put it in the Proc. Camb. Phil. Soc., which is just about to open a new series with a better page and style: I told him I was sure that you would not mind whether it was this or the Journal.

I have made quite a lot of red ink marks on it, and must get one of my pupils to make a fair copy.

The theory about my sister is picturesque and, though quite untrue, may well become the official version. Surely it is a little remarkable that (unless your friends are English) they should be able to identify the quotation [it is extremely familiar in English, but not in its context]. I should not have expected Browning to be known at all outside England. He was once 'all the rage,' but his day, even here, is over and he has no 'universal' appeal like Byron or Dickens. When I was a schoolboy, it was necessary to admire him.

Still, he said good things at times, and these two lines have passed into the language.

Do you know the two lines (and *only* two lines) of the great American poet *Ella Wheeler Wilcox*? Ask your wife!

Yours ever
G. H. Hardy

Still B. [Browning], at the lowest, was Mittag-Leffler to E. W. W.'s S. C. Mitra! With Shakespeare 100, Milton 73, Shelley 71, Tennyson 39, E.W.W. 2, I give him 27.

This letter was followed by a postcard where Hardy does identify the two lines of Ella Wheeler Wilcox (E.W.W.) to which he refers in the letter. Here is Hardy's postcard:

Trinity College
Cambridge

Laugh, and the world laughs with you.
Weep, and you weep alone.
                              (E.W.W.)
With:
'Duncan is in his grave,
After life's fitful fever he sleeps well' = 100
and the Browning quotation = 61
I give this 23. Otherwise E.W.W. = 0.07
                              G.H.H. (Hardy, 1990)

Here he generously gives Shakespeare a perfect score of 100 for the lines from Macbeth, and to Browning for the quotation alluded to, but not explicit in the letter, a mere 61. For the quoted lines from Wilcox he assigns 23 points but it is clear that he does not hold her other work in high regard, perhaps justifiable when one observes that her other famous lines are

> So many gods, so many creeds,
> So many paths that wind and wind,
> While just the art of being kind
> Is all the sad world needs

lines with a decided greeting card quality!

Assigning points and setting up rankings was not unusual for Hardy. In a ranking of mathematicians on the basis of natural mathematical talent he gave Ramanujan a perfect score of 100, followed by 80 for Hilbert and 30 for Littlewood. He modestly assigned a score of 25 to himself. (Kanigel 1991, 226)

The Plutocratic Years

In the 1930s the Pólyas had a chalet in the Swiss mountains, in Engelberg, during what Pólya referred to as their "plutocratic years" or their "capitalistic era." In the guest book for the chalet (the Hüttenbuch), Hardy's name appears for July 1931 and his sister's name appears for the month of August in 1931, 1935, and 1937. Hardy's sister, Gertrude Edith Hardy ["Miss G. E. Hardy" in the guest book], had lost an eye as a child, in a very unfortunate accident. It was apparently Hardy's fault. She was devoted to her brother, nevertheless, and to his memory after his death. Like her brother, she never married. She, too, was close to the Pólyas.

Visits to the Pólyas' chalet in Engelberg must have been mathematically stimulating to all those who visited there. The guest list through the 1930s included the names of some truly renowned mathematicians and physicists: Michel Plancherel, Hermann Weyl, Otto Blumenthal, Heinz Hopf, Mihály Fekete, Edmund Landau, Arthur Rosenthal, Issai Schur, Rolf Nevanlinna, Ernst Völm, A. E. Ingham, Oskar Perron, L. J. Mordell, Carl Ludwig Siegel, Louis Kollros, Albert Pfluger, Erich Hecke, John von Neumann, Heinrich Behnke, and physicists Fritz Zwicky and Wolfgang Pauli.

It was a visit by Hardy to the chalet in Engelberg that led to one of Pólya's stories about Hardy's well-known view of God. Titchmarsh in his obituary said, "Hardy always referred to God as his personal enemy. This was, of course, a joke, but there was something real behind it. He took

The chalet at Engelberg

his disbelief in the doctrines of religion more seriously than most people seemed to do. He would not enter a religious building, even for such a purpose as the election of a Warden of New College. The clause in the New College by-laws, enabling a fellow with a conscientious objection to being present in Chapel to send his vote to his scrutineers, was put in on his behalf." (Titchmarsh, 1949)

Pólya reported that Hardy liked the sunshine, but on a visit to Engelberg "it rained all the time, and as there was nothing else to do, we played bridge: Hardy, who was quite a good bridge player, my wife, myself, and a friend of mine, F. Gonseth, mathematician and philosopher. Yet after a while Gonseth had to leave, he had to catch a train. I was present as Hardy said to Gonseth: 'Please, when the train starts you open the window, you stick your head through the window, look up to the sky, and say in a loud voice: 'I am Hardy.' ' ... You have understood the underlying theory: When God thinks that Hardy has left, he will make good weather just to annoy Hardy." [1969, 4]

"Prove the Riemann Hypothesis"

When Hardy was not visiting the Pólyas in Switzerland or other friends in the sunnier parts of Europe, he would head for Denmark to visit his friend Harald Bohr, the mathematician brother of the famous physicist. "They had a set routine. First they sat down and talked, and then they went for a walk. As they sat down, they made up and wrote down an agenda. The first point of the agenda was always the same: 'Prove the Riemann Hypothesis.' As you are probably aware, this point was never carried out. Still, Hardy insisted that it should be written down each time." As another example of Hardy's ongoing feud with God, Pólya tells the following story: "Hardy stayed in Denmark with Bohr until the very end of the summer vacation, and when he was obliged to return to England to start his lectures there was only a very small boat available (there was no airplane traffic at that time). The North Sea can be pretty rough and the probability that such a small boat would sink was not exactly zero. Still, Hardy took the boat, but sent a postcard to Bohr: 'I proved the Riemann hypothesis. G.H. Hardy.' . . . If the boat sinks and Hardy drowns, everybody must believe that he has proved the Riemann hypothesis. Yet God would not let Hardy have such a great honor and so he will not let the boat sink." Obviously, since Hardy successfully reached England, this form of insurance was effective. [1969, 4]

Pólya and Hardy often took walks, discussing mathematics along the way, and when they came to a church, if necessary, Hardy would change places with Pólya so that Pólya would be walking closer to the church than Hardy, just in case God should strike out at Hardy with a lightning bolt.

Pólya was drawn to Hardy probably in part for the same reason that he was drawn to Fejér, and he admitted this. Just as Fejér's personality and, as viewed by some, eccentricities had drawn many people to him, similarly in England Hardy had this effect.

Pólya reported that Hardy "was strikingly good-looking, and very elegant when he put on a dinner jacket ...". [1969, 4] At the same time, Hardy could not stand to look at himself in a mirror. When he entered a hotel room, he draped towels over the mirrors so he would not have to look at himself. For the same reason he avoided having pictures taken of himself, yet Pólya had in his photograph collection a remarkable number of likenesses of Hardy, enough so that when Vladimir Drobot was looking through Pólya's picture album he remarked that Pólya had more pictures of Hardy than exist! [1987, 2, 121]

The Oxford-Cambridge experience for Pólya led to several papers co-authored with Hardy and Littlewood and one co-authored with Hardy and A. E. Ingham. Much of the work, however, resulted in the most

J. E. Littlewood

famous collaboration of Pólya with Hardy and Littlewood, the book *Inequalities,* which was published in 1934. R. Rado, writing in the *Collected Papers of G. H. Hardy,* said, "It can fairly be said that this book transformed the field of inequalities from a collection of isolated formulæ into a systematic discipline." (Hardy 1967, 2: 379) The book was intended originally for the Cambridge Tracts series, but the manuscript outgrew that format, running eventually to more than 300 pages. Since its publication it has never been out of print and remains a standard reference in the area of inequalities and applications of inequalities in analysis. The Cambridge University Press issued it in paperback in 1988. It has been cited in countless papers. *Inequalities* has been called "the first—and still the best—book on ... the subject," (Burkill 1995/96) and "the standard work on inequalities" (Korevaar 1977).

Of course, the collaboration on the book did not end in 1934. When a revised edition was due out in 1952, there was a flurry of correspondence on revisions between Pólya and Littlewood (Hardy had died in 1947). The exchange is fascinating if one recalls that the Hardy-Littlewood collaboration was carried out often by correspondence, even when both were living in Trinity College. Holding true to form, Littlewood's letters, though cordial, come right to the point, with little extraneous detail, but with direct remarks: he finds, for example, that one of Pólya's suggested proofs will be insufficient for the "stupid reader." The shortest letter of all, however, is a postcard, that reads in total: "All agreed. JEL."

TRINITY COLLEGE,
CAMBRIDGE.

All agreed.

JEL

The book contains (in somewhat more general form) a very ingenious proof of the arithmetic-geometric means inequality (Theorem 9 in the book, proof in section 4.2) where the first two terms of the Maclaurin series for e^x are used to provide a proof. Obviously considerable power is already present in the Maclaurin series for e^x, but using this, one can demonstrate the arithmetic-geometric means inequality quickly and easily.

Here's the proof: Given that $A = \dfrac{1}{n}\sum_{i=1}^{n} a_i$, $G = \left(\prod_{i=1}^{n} a_i\right)^{1/n}$, $e^x \geq 1 + x$,

$a_i \geq 0$, consider the inequalities $e^{\frac{a_i}{A}-1} \geq \dfrac{a_i}{A}$, $i = 1, 2, \ldots, n$. Then

$$\prod_{i=1}^{n} e^{\frac{a_i}{A}-1} \geq \prod_{i=1}^{n} \frac{a_i}{A} \quad \text{so} \quad e^{\sum_{i=1}^{n}\frac{a_i}{A}-n} \geq \frac{G^n}{A^n}, \; e^{\frac{nA}{A}-n} \geq \frac{G^n}{A^n}.$$ Then $A^n \geq G^n$ and

$A \geq G.$

When asked about his proof years later, Pólya replied that it was the best mathematics he had ever dreamt. He apparently did dream mathematics fairly often, but, like others, when he would get up in the morning and try to write it down, he usually found that it was either wrong or less original than it had seemed during the night. But this particular proof, dreamt sometime around 1910, was not disappointing in the cool light of day. On a later occasion he dreamt about "strong summability" of the Fourier series but noted later that the problem had already been considered and settled by Hardy and Littlewood. [1970, 2]

The academic year at Oxford and Cambridge was a stimulating one for Pólya. He continued to work, of course, on a number of problems that he had brought with him from Zürich, but the collaborations with Hardy, Littlewood, and Ingham clearly indicated the influence on him by the English school of analysts at the time.

Hardy went over Pólya's manuscripts carefully, particularly those submitted for publication to English journals. Pólya kept two bound sets of his own reprints prior to 1940; in one is included a manuscript for a paper of 1923 (before he went to England) extensively marked in Hardy's distinctive hand [1923, 8]. Dame Mary Cartwright has reported that in Hardy's rooms at Cambridge, he "kept papers methodically in piles on a large table in front of a window. I remember him saying 'a pile for Landau and a pile for Pólya.'" (Cartwright 1981)

Ramanujan's Notebook

While in England Pólya asked Hardy if he could see Ramanujan's first notebook and Hardy lent it to him. Bruce Berndt reports that "A day or so later, Pólya returned the notebook in a state of panic. He said that as long as he held on to the notebook, he would continue to try to prove the formulas in it. The notebooks were so fascinating that Pólya was afraid that if he kept them, he would never again prove any result of his own." (Kolata, 1987) It is regrettable that Pólya's visit to England came a few years late to meet Ramanujan. In the film about Ramanujan shown on the television series *Nova*, Béla Bollobás commented that only Ramanujan could have conjectured the explicit formula for the partition function $p(n)$, but once conjectured there were several people who could have proved it. Hardy did, but Bollobás said that Pólya was one of those who could have.

The collaboration with Hardy, as indicated earlier, extended to a collaboration with A. E. Ingham, author of the important Cambridge Tract on the distribution of primes. Ingham was a fellow of King's College, Cambridge and a major contributor in the area of analytic number theory. His association with Pólya went beyond the collaboration that occurred during the year that Pólya spent in England. A bit earlier, in 1919, Pólya had conjectured that among the numbers 1, 2, ..., n, there are always more with an odd total number of prime factors than with an even number, or at least as many [1919, 2]. In this case if a prime appears in a factorization more than once it is counted separately each time it appears. Thus, for this discussion $12 = 2^2 \cdot 3$ would have three prime factors. It is in many ways a plausible conjecture since the primes themselves have an odd number of prime factors, and composite numbers have to follow the primes that go into their factorization, in a sense. It would also seem equally likely that a composite number is a product of an even number of prime factors or an odd number of prime factors. Hence, since the prime factors precede the composite numbers of which they are factors, it would seem that the number of those with an odd number of prime factors up to n would be larger than the number with an even number of prime factors. It turns out that this conjecture, if true, would imply the truth of the Riemann hypothesis. [1919, 2] (For a more formal statement of the conjecture, see Appendix III.) The conjecture languished for many years. On Pólya's birthday (by coincidence) in 1931, S. Chowla wrote to Pólya from Delhi that he had been told of the conjecture by André Weil (who had heard it from Pólya) and he had been looking at the problem. He described some partial results, acknowledging that a proof would imply the Riemann hypothesis. (Chowla, 1931) In January of 1932 Pólya wrote to Chowla expressing doubt about some of Chowla's calculations, particularly in the cases of 585 and 586 — a long way from the counterexample that people were looking for. (G. Pólya 1932) "In 1942 Ingham showed that it would imply much more, namely, that the ordinates of the zeros of the ζ-function would be connected by infinitely many linear relations with integer coefficients. He proved a more specific result, involving only a finite number of the zeros, and so opened up the possibility of disproving the conjecture by computation." (Burkill 1968) This was eventually done in 1958 by a former research student of Ingham's, C. B. Haselgrove, who showed that the conjecture fails for some $x < e^{832}$. (Haselgrove 1958) "Ingham's method is not limited to the investigation of Pólya's conjecture and is one of great generality." (Davenport 1967) Haselgrove disproved the conjecture by showing the existence of infinitely many x's for which it fails, but without demonstrating a single one! R. Sherman Lehman found in 1960 a counterexample to Pólya's conjecture, namely, 906,180,359, but this counterexample was not the least such. (Lehman 1960) That was found by Minoru Tanaka in 1980; it is almost as large as Lehman's example: 906,150,257. (Tanaka 1980)

A Letter of Recommendation

Some years after Pólya went back to Zürich from England, Hardy was asked to write a letter of recommendation in support of a promotion for Pólya at the ETH. This gives some details concerning Hardy's opinion of Pólya's work, including their own collaboration:

> I have a very high opinion of Pólya both as a mathematician and as a person, an opinion which was greatly strengthened during his stay here. I feel too that it is very probable that the best has not yet been seen of him; that he has, perhaps got a little discouraged during recent years; and that the encouragement derived from an improvement in his position would be very likely to stimulate him to even better work. He always seemed to feel that his leisure was scanty, and that he dared not risk squandering any large part of it on a difficult problem that might lead nowhere; and of course,

when a fine mathematician begins to feel like that, there is always a serious danger that he may not do justice to his powers.

What is peculiarly characteristic of Pólya is that he has very beautiful ideas—sometimes, of course (as in the classical case of the 'Carlson-Pólya' or the 'Pólya-Carlson' Theorem—where all the *ideas* were his), his ideas have run away from his executive powers! But all his work in that field is singularly beautiful. I have of course collaborated with him twice on a considerable scale, and I was always astonished by the intensity of his aesthetic sense. Where L. [Littlewood] and I were content to get at the results *somehow,* and have it at that, he simply could not rest until he had got the whole corpus of theorems into perfect aesthetic shape (it is because of that, of course, that L. and I were so determined to get him as a collaborator in our proposed "inequality" Tract). Apart from that, he once rescued L. and me from a month of sheer despair. In our triangular paper (*Proc. L. M. S.* 25) it was he who provided the happy idea that finally led to success. The situation then (this is naturally a piece of 'unofficial' information) was that L. and I had formulated, and thought we have proved, a collection of theorems of which we foresaw all sorts of applications. At the last moment— when part of our work stood already in print—we found that our main argument was entirely and in principle fallacious. We were altogether unable to put it right, and it was P's intuition—a quite new 'maximal' idea—which saved the situation and on which the whole investigation was ultimately based.

But this is a small and special affair. In general I would say that I have read many of P's memoirs, and never yet found one which did not show real insight and originality, or some really interesting theorem expressed in a beautiful way.

When P. came here, he was an immense success in every way. I happened to have several quite good pupils at the time, and you can imagine what a difference it made to have a first rate analyst in the place; especially one with that 'algebraical' touch that is becoming more and more important in analysis. He was extraordinarily good in helping me with them, though he was under no kind of obligation to do so, both in taking part in my lectures and classes and in providing my pupils with ideas. You can see something of the results still, if you turn over the pages of the new *London Math. Soc. Journal,* and observe the number of interesting little papers in the 'inequality' field. I am sure that if Zürich can provide him at last with a position commensurable with his abilities, it will be doing a very great service both to its own reputation and to science.

> With all kind regards, I am
> yours very sincerely
> G. H. Hardy

7
The United States—
The First Visit

While still at Oxford, Pólya received an invitation from E. Diencs (written partly in Hungarian), who was at the University College of Swansea in Wales, to give a series of lectures at Swansea. These were scheduled for the spring of 1925, well after the five week break the Pólyas spent in London around Christmas time. Beginning in April of 1925 they moved to Cambridge where Pólya worked at Trinity College. The year in England was surely for both a wonderful experience. Compared to the rather frugal time spent in Zürich when Pólya was a mere *Privatdozent,* life on a Rockefeller Fellowship at Oxford and Cambridge must have seemed rather opulent. Stella Pólya was particularly impressed by the elegant touches, like the menu, one of which she kept, dated June 8, 1925: Eclairs Jacqueline/Salmon Chambord/Roast lamb, Roast duckling, peas, French beans, new sauté potatoes/compote of Greengages and cream, apricot cream ice/croûte Hollandaise. It's a life one could get used to.

In the years that followed Pólya's return to Zürich, there were further invitations to spend time as a visiting lecturer, ranging from a few days, allowing a series of lectures, to three months. In 1929 he visited the University of Paris, and in 1932 the University of Göttingen.

It was probably about the time of this latter visit that Pólya became aware of the whimsical interpretation of the reproduction of Titian's painting, "Sacred and Profane Love," hanging on the wall of the Mathematical Institute at Göttingen in the twenties. He was to write of it later for

81

Titian's Sacred and Profane Love (Amor sacro e profano)

Springer. First, one has to know that the phrase "pure and applied mathematics" in German is *"reine und angewandte Mathematik."* Profane Love, the unclothed figure on the right, corresponds to *reine Mathematik,* since the clothed figure on the left is applied mathematics, being *"Angewandte"* (*gewand* refers to a dress or, more generally, clothing). It has been observed, however, that there is hope that Pure Mathematics might turn out to be applied since here even Pure Mathematics is at least minimally clothed. The putto running his arm through the water represents hydrodynamics and on the sarcophagus the figure bludgeoning another with a club represents mathematical pedagogy! Pólya enjoyed telling this story, exhibiting his delight in a good joke. Titian would have been mystified by this interpretation of the picture, however; the clothed figure was really Laura Bagaretto of Padua, the unclothed was Venus—clearly the same woman—and the picture was commissioned for Bagaretto's wedding. (Rowland, 1999)

Also in 1932 Pólya received word from the Rockefeller Foundation that the directors were again willing to support his research with a special fellowship in 1933 to visit Princeton University for three months and, subsequently, Stanford University for three months. The grant was probably generous for the time: $2,600. The Foundation did, however, set out some detailed and restrictive conditions to its Fellows. Here are a few examples from the regulations:

> No regular vacation is permitted. These fellowships are awarded for intensive work in specific fields, and for specific periods. It is felt that needed periods of recreation should be taken before beginning, or after finishing the fellowship.

Upon arrival in the United States, a Fellow should not cable home at the expense of the Foundation.

It is requested that Fellows refrain from using the name of the Foundation on visiting cards or personal stationery.

The Foundation kindly offered, however, to arrange for passage on the United States Line's Manhattan, which would sail for the United States on March 9, 1933. (The Pólyas ended up sailing on the S. S. Bruner.) The cost, which would come out of the $2,600, was $150 per person. So on April 13 of that year he was appointed special lecturer in mathematics at Princeton for the second term of the 1932–33 academic year.

The Three Nicest Mathematicians

Princeton had then, as it has now, a splendid department of mathematics. It is not entirely clear, however, why Pólya chose to visit Princeton since the mathematical tastes there were, for the most part, not close to his own. There is no evidence that anyone at Princeton had any great influence on him. (The Institute for Advanced Study was only beginning to function in 1933, having been chartered in 1930.) He did find Oswald Veblen a most congenial colleague and quoted Hardy, who had highly recommended Veblen, as saying that Veblen had "all the American, all the British, all the Scandinavian virtues." Pólya was later to describe Veblen as one of the "nicest" mathematicians he had met. But Veblen did not make his later list of the three "nicest" mathematicians he had ever known: Harald Bohr, Heinz Hopf and Karl Löwner. This gave him an opportunity to list the three most unpleasant mathematicians he had ever known and he did—but that list is conveniently lost.

Of course, one professor at Princeton he knew quite well—his former student, John von Neumann, who had gone to Princeton in 1930. While a student at the ETH, von Neumann had taken an advanced course from Pólya. Ulam in writing about von Neumann many years later recalled von Neumann's positive experience in studying with Pólya and Weyl at the ETH. (Ulam 1958) However, the experience may have been remembered more fondly by von Neumann than by Pólya, who wrote: "[von Neumann] is the only student of mine I was ever intimidated by. He was so quick.… I came to a certain theorem, and I said it is not proved and it may be difficult. Von Neumann didn't say anything but after five minutes he raised his hand. When I called on him he went to the blackboard

*Uspensky, Pólya,
and Blichfeldt*

and proceeded to write down the proof. After that I was afraid of von Neumann." [1987, 2: 154]

At the invitation of H. F. Blichfeldt Pólya spent the summer quarter of 1933 at Stanford University, where he was given an allowance of $200 for living expenses by the University, beyond his Rockefeller stipend. By the time of his departure in August he had made an impression on the campus. Elliot Meers of the Summer Quarter Office informed him that his "colleagues and students speak of your work in highest terms." Unfortunately Professor Meers went on to say that it was "doubly unfortunate... that the arrangement with the Rockefeller Foundation does not permit us to offer you, as you are doubtless aware, any financial remuneration." Nevertheless the short stay at Stanford was to be influential in choices that he would have to make a few years later in Zürich.

The stay in America was valuable in that it allowed Pólya to make contact with G. D. Birkhoff at Harvard as well as with Veblen at Princeton, two American mathematicians at that time with real international stature (of course, there were others: W. F. Osgood, L. E. Dickson, J. W. Alexander, A. A. Albert, for example, but Birkhoff and Veblen were probably best known), as well as with the group of mathematicians at Stanford. The department head at Stanford at that time was Blichfeldt, who had been a student of Sophus Lie's in Leipzig. Other mathematicians at Stanford included W. A. Manning, the group theorist, and J. V. Uspensky, who worked in number theory and probability theory. (Uspensky had been the teacher of Vinogradov in Russia.) The direction of the Department did not tilt in the direction of analysis and applied mathematics till several years later with the arrival of Szegő who became Executive Head. (Royden 1989)

Pólya was not the only distinguished visitor from Europe to visit Stanford in the 1930s. Blichfeldt had done significant work in the geometry of numbers and in group theory. He therefore had good contacts in Europe, so he could attract outstanding visitors. Landau had come for the summer of 1931, right after Harald Bohr had spent the full academic year 1930–31. Szegő had visited in the summer of 1935, just after leaving Königsberg. And four years later Emil Artin visited. (He had been offered a position at Stanford in 1936, but the Nazis had not allowed him to accept it; a year later they forced him to leave Germany.) And these were not the only distinguished summer visitors: American visitors included Gordon Whyburn, Marshall Stone (the same summer as Pólya), Dunham Jackson, D. H. Lehmer, Rudolph Langer and Lawrence Graves. (Alexanderson and Klosinski 1988)

If Stanford was impressed by Pólya, it's also true that Pólya was impressed by Stanford.

~ 8 ~

Swiss Citizenship

Between visits to Cambridge, Oxford, Paris, and other centers of mathematical life, Pólya was still spending plenty of time in Zürich, teaching and doing research. One place that Pólya did not visit, however, was Hungary. This has an explanation. When he was still in school he had suffered a serious soccer injury: his leg became badly infected and it required surgery. This injury kept him out of the Hungarian army during the early years of the First World War, and by the time the war progressed and he might have felt a duty to return and serve in the army, he had fallen under the influence of Bertrand Russell, a well-known pacifist. Pólya thereafter refused to return to Hungary to fight and for many years after that he felt that he was unable to return safely even to visit his family who had remained there. He stayed in Switzerland and became a Swiss citizen.

The problems of visiting his family were alleviated by the fact that his mother, Anna, was able to visit Zürich from time to time. Stella Pólya never seemed anxious to talk about her mother-in-law, but there were hints that Anna Pólya was a rather imperious woman and surely would not have been the image of the perfect mother-in-law in the eyes of most brides. Relations could be a bit strained. Anna had decreed after the death of her husband that one of her daughters could not marry but would instead have to devote her life to taking care of her mother. This task fell to Ilona who did indeed never marry.

In 1924, the Pólyas went to Vienna in order to see Pólya's brother Jenő and his family. Anna and the two sisters, Ilona and Flóra, also came

One of Pólya's needlework designs for his niece Susi

so it provided a setting for a pleasant reunion. Pólya remained close to his nieces and nephews but especially close to John, Jenő's oldest son, possibly because John spent his summer holidays with the Pólyas in Zürich in 1928. John then came to Zürich to attend the ETH in 1931. Pólya made up needlework patterns for his niece, Susi, some easily seen to be figures of people and such, but others suggestive of fractals, raising the question of his having been influenced by Julia sets, long before they became so widely known.

Pólya's first Ph.D. student at the ETH was Walter Saxer in 1923. The following year he had two Ph.D. students receive their degrees: Alfred Aeppli and Florian Eggenberger, followed by Emil Schwengeler in 1925, Fritz Gassmann in 1926, August Stoll in 1930, Reinwald Jungen in 1931, Ernst Boller and Gottfried Grimm in 1932, Eduard Benz and Albert Pfluger in 1935, Hermann Muggli and Alice Roth in 1938, and Albert Edrei in 1939. Three of these later joined the faculty of the ETH: Saxer,

Gassman and Pfluger. In fact, Pfluger replaced Pólya when Pólya came to the United States in 1940. In addition, over these years, Pólya was "*Korreferent*" (secondary advisor) for the dissertations of Hans Odermatt in 1926, Wilhelm Machler in 1932, Victor Junod in 1933, Hans Arthur Meyer and Egan Moecklin in 1934, and Hans Albert Einstein and James J. Stoker in 1936. Hans Albert Einstein was the younger son of *the* Albert Einstein, but more about him later. Stoker was to become a distinguished professor at New York University. Beyond this, Pólya assisted in advising the Ph.D. dissertations of Egon Lindwart, with Edmund Landau, in 1914 at Göttingen, and of Nikolaus Kritikos, with Rudolf Fueter, in 1920 at the University of Zürich.

One of the most interesting of these students was Eggenberger and it is with him that Pólya wrote a series of papers in probability describing what later was to be known as the Pólya-Eggenberger scheme or the Pólya urn scheme. [1923, 7; 1928, 6; 1931, 1] Assume that we have an urn containing r red balls and b black balls, and one ball is drawn at a

Pólya in the late 1930s

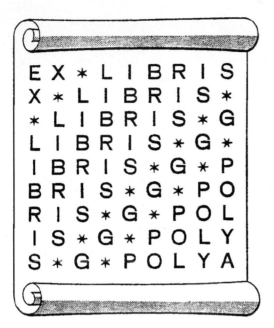

```
E X * L I B R I S
X * L I B R I S *
* L I B R I S * G
L I B R I S * G *
I B R I S * G * P
B R I S * G * P O
R I S * G * P O L
I S * G * P O L Y
S * G * P O L Y A
```

Pólya's bookplate

time. If a red ball is drawn, then $1 + c$ balls of that color are added to the urn; similarly for a black ball. Success, if that's described as picking, say, a red ball, leads to more success because red balls are added to the urn after a red ball is drawn, so the probability of drawing a red ball becomes higher. This provides a model for the spread of contagious disease. Jerzy Neyman described the scheme years later, saying that it was a mathematical device that was developed originally in order to understand the phenomenon of an influenza epidemic. The problem is one for which Pólya is well-known and it has had widespread impact. The idea sounds simple, but the consequences are far-reaching. Pólya reported on these results concerning contagion in March of 1929 at the Institut Henri Poincaré, where he spoke at the invitation of Émile Borel.

In 1928 the Pólyas had attended the International Congress of Mathematicians (ICM) in Bologna and Florence. This was the first ICM after the First World War to be attended by German mathematicians. At the 1920 Congress in Strasbourg, mathematicians from the Central Powers had not been invited and this was particularly provocative since Strasbourg lay in Alsace, in territory ceded to France by Germany under terms set by the 1918 Armistice. Again, in 1924, at the Congress in Toronto, Central Powers mathematicians were not invited. So Bologna was host to the first truly international Congress since the one held in Cambridge, England, in 1912.

Pólya gave two talks in Bologna, one in the analysis section, on singular points of series, and one in the probability session, on Gaussian distributions. He also was one of the "Presidenti de Sezione," presiding at a September 4 session on analysis. (Looking back one is surprised to note in the Proceedings that one of the Presidents of the Congress was Benito Mussolini! But this was 1928.)

Four years later, when the Congress met in Zürich, Pólya did not speak to the Congress, but he was, not surprisingly, a member of the Organizing Committee. And in Oslo in 1936 Pólya was one of the Presidents for the General Conferences, distinguished company that included Gaston Julia, Harald Bohr, E. T. Whittaker, Solomon Lefschetz, Torsten Carleman, Wasław Sierpiński, among others.

Jacques Hadamard at the Zürich Congress in 1932

Somewhat earlier, in 1920, Pólya, while studying normal distributions, coined the term "central limit theorem," a phrase first given in German (*zentraler Grenzwertsatz*), naturally. This arose in his work that led to the Pólya characteristic functions. To be sure, some of his work in probability was scarcely distinguishable from work in analysis because the topics considered included Fourier transforms and his work on roots of equations: the problem of roots of equations with random coefficients providing the relation of analysis to probability. The urn scheme led to something later called the Pólya distribution. Pólya also investigated a sufficient condition for a real function to be a characteristic function of a probability distribution, now called the Pólya criterion. [1949, 3] So with random walk, as described earlier, and work on these other probabilistic topics, one can easily see that Pólya's influence on probability was both broad and deep. For technical details of these contributions see Appendix 1.

Pólya was interested, for example, in the problem of deducing properties of analytic functions from properties of the coefficients of their power series. Pólya proved a conjecture of Fatou—that the circle of convergence of a series is usually a natural boundary, i.e., by changing signs of the coefficients one can make the sum of the series non-continuable. [1916, 1; 1929, 1] About this time he also proved an extension of the Fabry gap theorem, that a series is noncontinuable if the density of zero coefficients is one. Pólya showed that if the density is less than one but still positive, the singular points still occur but less frequently. This generalization is known as the Pólya gap theorem; he further extended the theorem in various directions. [1923, 4; 1929, 1]

On the question of Pólya's contributions to the relationship between the properties of the coefficients of the expansion of a function and properties of the function, Neyman said that from Pólya's Paris lecture in 1929, he concluded Pólya's work had somehow been inspired by that of Hadamard, who did work in this area, but Neyman said that after he read Pólya's collected papers on the subject, he wondered whether "the situation was the reverse, with Pólya's results inspiring an effort by Hadamard." (Neyman 1977B) (This is probably unlikely, given the difference in their ages. Hadamard was born in 1865 and had proved the prime number theorem by the time Pólya was nine years old!)

Another problem that Pólya worked on in the early years in Zürich was the question of the connection between properties of an integral function and properties of the sets of zeros of polynomials that approximate the integral function. These questions led to the development of functions that are now referred to as the Pólya-Schur functions or the Laguerre-Pólya functions.

About this time Pólya came up with a theorem, related to a theorem of Chebyshev's, which became a favorite of Erdős's, both for its surprising conclusion and for the elegance of its proof. [1928, 5] It is included in the compilation of Erdős's problems in The Book. (Aigner and Ziegler, 1998, 110–116) It must be explained that Erdős claimed that God keeps a book with the best proofs, so if a proof is as clean and elegant as a proof can be, it must appear in God's Book. Pólya's theorem (and proof) that made The Book is this: if we have a complex polynomial of degree $n \geq 1$ with leading coefficient 1

$$f(z) = z^n + b_{n-1}z^{n-1} + \cdots + b_0$$

then, associated with $f(z)$ is the set $C = \{z \in \mathbb{C} : |f(z)| \leq 2\}$. C is the set of points mapped under f into the circle of radius 2 around the origin in the complex plane. Then if one takes any line L in the complex plane and considers the orthogonal projection C_L of the set C onto L, the total length of any such projection never exceeds 4. This holds for any polynomial with leading coefficient 1.

––––––––––

During the early years in Zürich Pólya was doing major work in real and complex analysis and probability, while at the same time, with Szegő, he was producing the *Aufgaben und Lehrsätze*.

As early as 1919 Pólya was considering problems in number theory, albeit problems related to questions in analysis. It was during that year that he stated his Pólya conjecture on the total number of prime factors, described earlier, and disproved a conjecture of Fekete on polynomials with Legendre symbols as coefficients. [1919, 2]

In later years Pólya was asked how he happened to work in so many different fields and he replied that it was in part due to his interest in problems. He said, "I was partly influenced by my teachers and by the mathematical fashion of that time. Later I was influenced by my interest in discovery. I looked at a few questions just to find out how you handle this kind of question." (Albers and Alexanderson 1985, 244)

Pólya—Like a Bear

Earlier on, Hardy had criticized Pólya for moving from problem to problem. Pólya described Hardy's attitude as follows: "In working with Hardy,

I once had an idea of which he approved. But afterwards I did not work sufficiently hard to carry out that idea, and Hardy disapproved. He did not tell me so, of course, yet it came out when he visited a zoological garden in Sweden with Marcel Riesz. In a cage there was a bear. The cage had a gate, and on the gate there was a lock. The bear sniffed at the lock, hit it with his paw, then he growled a little, turned around and walked away. 'He is like Pólya,' said Hardy. 'He has excellent ideas, but does not carry them out.'" [1969, 4]

This quality that Pólya had of finding good problems in many areas and then turning them into good mathematics was one of his strengths. His questions were good, but his answers were also good in that they were models of clarity. H. L. Royden emphasized that Pólya was not only a mathematician, he was also a teacher. For him it was not sufficient to solve a problem—he had to study it until he saw the solution clearly so that he could put it in a form easily accessible. (Royden, 1977) An example of this is a paper that he wrote estimating electrostatic capacity. In this paper Pólya discussed the problem of minimizing capacity and in particular thermal conductance, so he gave the example of a cat preparing itself for sleeping through a long, cold winter night: "He pulls in his legs, curls up, and, in short, makes his body as spherical as possible." In doing this the cat demonstrates knowledge of the following important theorem: "Of all solids of a given volume, the sphere has the minimum capacity." [1947, 1] This kind of homely description, similar to his describing his result in two-dimensional random walk as "all roads leading to Rome," was a specialty of Pólya's and one appreciated by his students.

Be As Concrete as Possible

Pólya's mathematical interests were in some ways close to Euler's. Like Euler, he preferred mathematics that was based on specific examples. He was not comfortable with vast theoretical constructs of the sort that one finds in certain branches of algebra and topology. He was drawn to problems that arose naturally out of physical problems, but also from many other subjects as well. In support of this attitude, Pólya said that mathematics is the most abstract of the sciences, so in teaching it (or writing it) one must be as concrete as possible. Of the two kinds of problems described by Poincaré, those that pose themselves ("qui se posent") and those that people pose ("qu'on se pose"), Pólya obviously would choose the former. His Ph.D.

student and successor at the ETH, Albert Pfluger, described Pólya's tastes as follows: "Pólya was attracted by problems originating in physical sciences and engineering, and many of his mathematical developments were motivated by such problems. Characteristic is his special liking for the concrete, but typical, particular case by which the general idea can be seized and comprehended or a general method can be verified." (Pfluger 1977)

In support of his admonition to consider concrete problems, Pólya cited what he thought would be Littlewood's reaction: "Relatively concrete problems, such as the proof of the Riemann hypothesis, are less in vogue nowadays, for reasons partly good and partly bad—'Mostly bad,' Littlewood would interject if he were present." [1969, 4]

Symmetry and Escher

There is another key to Pólya's breadth of tastes, however. He was acutely aware of the fact that he was teaching students who most often were not mathematics majors—that is, they were not students who would study mathematics for its own sake. At the ETH he was teaching students in engineering and physics, subjects that he had studied extensively himself, but also students of architecture, chemistry, and even forestry. In part this was the result of the emphasis at the ETH on technology and applied science. While doing challenging and fundamental work in mathematics, he still found time to visit the Architecture Library, for example, where he found architectural ornaments to be of great interest. This led to his often cited paper on the seventeen plane symmetries. [1924,1] Pólya put it this way, "The architecture department had an interesting library. I studied there architectural ornaments and that led to one of my papers: about the analogy of the symmetries of crystals in the plane. I think one point in it is new. I illustrated the seventeen-plane symmetries with ornaments. I must confess the architects were not so much interested in it. But the professor of mineralogy, [Paul] Niggli, was very much interested in it and he wrote a parallel paper. (Niggli, 1924) The seventeen symmetries had been discovered earlier by E. S. Fedorov, but that was apparently not known to Pólya at the time. Pólya, however, went further in providing illustrative examples for each of the symmetries, many taken from Moorish sources, the Alhambra, in particular. Four, however, had to be "invented" by Pólya to demonstrate symmetries that he had proved must exist by looking at the symmetry groups.

This paper, when called to the attention of the Dutch graphic artist M. C. Escher, prompted Escher to produce more sophisticated work. Knowing as we do now the ingenious work of Escher, we can almost see some of his creations when we look at the illustrations Pólya provided. Amazingly enough, Pólya had shown a certain prescience when he remarked at the end of his 1924 paper that this mathematical study of ornaments might be of interest to artists. There was some delay—Escher did not contact Pólya till 1937. Pólya responded to Escher's initial letter and Escher sent him a drawing. Unfortunately these seem to have been lost when the Pólyas left Zürich. For details see (Schattschneider 1987), Schattschneider's Appendix 4 to this volume, or her classic book, *Visions of Symmetry* (Schattschneider 1990, pp. 22–26).

Pólya's Enumeration Theorem

This early interest in symmetry culminated in 1935 in one of Pólya's most renowned contributions, the famous work on the combinatorial enumeration of groups, graphs, and chemical compounds. [1935, 4] His teaching of chemists contributed to his interest in this field, and the paper was motivated initially by the solution of the problem of finding the appropriate model for the benzene molecule. Possibilities were the regular hexagon, the regular octahedron, and the right triangular prism. As it turns out, the regular hexagon was the correct model, the so-called benzene ring, suggested by Kekulé.

These studies in chemistry and in crystallography motivated the question of how many labelings of a geometric figure are equivalent under rotations in space. Or, put differently, if the vertices of a polyhedron are assigned different colors, possibly representing atoms of different elements in a molecule, for example, one can easily count the number of ways one can assign a fixed number of colors to these vertices. But not all of these colorings will be "different" since one coloring may be seen to be equivalent to another if the polyhedron is rotated about an axis of symmetry. Coloring all the vertices of a cube but one red and coloring the remaining vertex blue is an example. The blue vertex might appear on the top right front, or at the back, or somewhere on the bottom of the cube, but we would not like to think of these as "different" since they look different only depending on the direction from which one is looking at the cube. Counting the number of colorings that are "different" is a considerably harder problem than just assigning colors in all possible ways and counting them. This is the problem addressed by Pólya's theorem. The theorem has many ramifications. Later on, Pólya said, "Again the chemists were not too much interested in it but Niggli was very much interested. He even used the ideas of my paper in his [class] which was obligatory for chemists and other people and he included some of it in his textbook."

As it turns out, the result had been anticipated by J. H. Redfield who had published an earlier paper on the subject in the *American Journal of Mathematics.* (Redfield 1927) Unfortunately the notation and the methods used were sufficiently obscure that the paper had not attracted any attention. And it was surely not known to Pólya. He saw the problem for what it really was, a generalization of Burnside's lemma in group theory, and he further saw the applications outside mathematics. This failure of the mathematical community to recognize Redfield's contribution to the problem is the subject of some verse by Blanche Descartes (pseud.).[1] (Berge 1968, 7)

"ENUMERATIONAL"

Pólya had a theorem
(Which Redfield proved of old).
What secrets sought by graphmen
Whereby that theorem told!

So Pólya counted finite trees
(As Redfield did before).
"Their number is exactly such,
and not a seedling more."

Harary counted finite graphs
(Like Redfield, long ago),
And pointed out how very much
To Pólya's work we owe.

And Read piled graph on graph on graph
(Which is what Redfield did).
So numbering the graphic world
That nothing could be hid.

---

[1] Blanche Descartes is assumed by Richard K. Guy to be William Tutte but there are others who have contributed under this pseudonym.

Then hail, Harary, Pólya, Read,
Who taught us graphic lore,
And spare a thought for Redfield too,
Who went too long before.

— Blanche Descartes

The enumeration theorem has been generalized in various directions. For an extensive account, read R. C. Read's Appendix 5 or his essay accompanying the English translation of the original paper. [1987, 1] Subsequent work not included there has been done for infinite groups by Robert A. Bekes. (Bekes, 1992)

The celebrated paper describing the enumeration theorem now bearing Pólya's name was not his first foray into combinatorial problems. In 1918 he had published a paper [1918, 6] on doubly-periodic solutions of the n-queen problem, placing n chess queens on an $n \times n$ board where no pair of queens can attack one another. The work is described in detail by David Klarner in [1984, 2, 4: 613–614].

An article on the distribution of plants by Paul Jaccard, a biologist at the Technical Institute in Lausanne, prompted Pólya to write an article on that subject [1930, 5], and a result related to Pólya's famous urn scheme was the subject of Hans Albert Einstein's dissertation in 1936, on the movement of silt in rivers. (Hans Albert Einstein was, as mentioned earlier, the physicist's younger son and, in part, Pólya's Ph.D. student at the ETH. Einstein's family lived just a few doors down from the Pólyas on the Büchnerstrasse; Albert Einstein had, however, left for Berlin in 1914.) Pólya went on to write on problems of silt movement himself in two papers. [1937, 1; 1938, 2] H. A. Einstein eventually became a professor of hydraulic engineering at the University of California, Berkeley. All of these moves into subjects outside mathematics resulted in interesting mathematical techniques. And this broadened Pólya's range of mathematical and problem solving skills.

In 1936 Pólya spoke at Hadamard's renowned seminar at the Hadamard jubilee in Paris. This seminar, begun by Hadamard in 1913 and interrupted by World War I, continued for 20 years. Pólya had been one of the first of a long line of illustrious speakers. (Maz'ya and Shaposhnikova 1998)

Also in 1936 the Pólyas headed north to Oslo to attend the last prewar International Congress of Mathematicians. It was not their first trip to the northernmost part of Europe. They had earlier visited the Nevanlinnas in Finland where they were entertained at the Nevanlinnas's

*Pólya and Dame Mary
Cartwright at the Oslo
Congress in 1936*

summer home in the country, where one had to be taken by horse and wagon from the closest railway station. That was their first encounter with the sauna and the practice left a lasting impression.

Pólya spoke in Oslo on his recently discovered enumeration theorem. Though there were ominous signs of trouble in Europe, the Oslo Congress drew from among Europe's (and the world's) best. By 1936 a number of German mathematicians had already emigrated, so in Oslo there was a mix of those who were still living in Germany and those who had already left. Only five mathematicians from Italy came, and there was no official delegation from Italy. The turmoil resulted in poor attendance, but in spite of that, the Congress was a notable one—the first Fields Medals were awarded there, for one thing. Also there exists a motion picture film of conversations taking place at the Oslo Congress, possibly another first. (A copy of the film is in the Stanford University Archives.) (See [1987, 2] for interesting pictures from the Congress.)

The Pólyas recalled with great pleasure the cruise arranged as part of the Congress in Oslo. (It seems there is always a boat ride as one of the recreational activities at an International Congress. Some have suggested that proximity to a body of water is the deciding criterion in selecting a site for a congress of mathematicians!)

Dinner on the boat cruise at the Oslo Congress (Carathéodory on the left looking over shoulder, Fréchet with gray hair and mustache looking at the camera, Hermann Weyl in the background to the right of the column, Norbert Wiener on extreme right)

In Oslo the reception at the Royal Palace was not, however, a great success. The mathematicians arrived ravenous — the Congress took place during the worldwide Depression of the 1930s — and the meager, but under normal conditions adequate, food supply disappeared within minutes. Mrs. Pólya was scandalized by what she viewed as a demonstration of bad manners. The King's reaction is not recorded.

~ 9 ~
Stanford

By 1940 it was clear that Pólya was looking for another job. He put to-gether a resumé on which he listed three books, the *Aufgaben und Lehrsätze*, *Inequalities*, and an early book that he had written on probability and statistics for the Abderhalden's *Handbuch der biologischen Arbeitsmethoden* (1925), as well as what he regarded as his principal papers. He broke the subject matter areas down into the theory of functions, the theory of numbers, arithmetical properties of analytic functions, the calculus of probability, location of roots, and "various subjects" (which included his famous work in combinatorics as well as his work in applied mathemat-ics, pedagogy and logic).

Life was pleasant in Zürich—the Pólyas had a house at No. 4 on the Dunantstrasse, and they had their chalet at Engelberg. The Zürich house was sufficiently large to include a maid's quarters and Pólya's status as a full professor at the ETH would have guaranteed the couple a comfort-able life. But Europe was not a comfortable place to be in 1940. Events had become more and more threatening since the early 1930s. Switzer-land was safe, but perhaps not entirely safe. In 1940 there was a report, later determined to be untrue, that the Nazis were going to invade Switzer-land. Pólya was told he would have to proceed to a preassigned hiding place. This incident removed any remaining doubt that the Pólyas should proceed with their plans to leave. Pólya's mother had died the previous year in Budapest, so leaving Europe posed fewer personal problems.

As early as 1934 Pólya had written to J. D. Tamarkin at Brown to inter-cede on behalf of Szegő by enlisting Tamarkin's help in finding a suitable dr. Pólya Jenő, özv. dr. Kemény Bélánc sz. Pólya Flóra dr. Pólya György gyermekei úgy a maguk, mint az alulírottak és az összes rokonok nevében is mély fájdalomtól megtört szívvel jelentik, hogy a forrón szeretett, felejthetetlen, jó édesanya, anyós, nagyanya, dédanya, testvér és rokon

Özv. dr. Pólya Jakabné
sz. Deutsch Anna

a Generali Biztositó Társaság nchai jogtanácsosának özvegye

folyó hó 24-én d. e. 11 órakor, életének 87-ik évében, hosszú szenvedés után, az Úrban csendesen elhunyt.

Drága halottunk földi maradványa f. hó 26-án d. e. 11 órakor fog a Kerepesi-úti melletti temető halottasházában a róm. kat egyház szertartása szerint beszenteltetni és ugyanazon temetőben örök nyugalomra helyeztetni.

Az engesztelő szentmise áldozatot lelkének üdvéért f. hó 27-én reggel 7 órakor a Ferencvárosi alsór Margit-körúti templomában fogjuk a Mindenhatónak bemutattatni.

Budapest, 1939. november 24.

Áldás és béke drága poraira!

dr. Pólya Jenőné sz. Borbás Lili
dr. Pólya Györgyné sz. Weber Stella
menyei

dr. Pólya János
Lányi Györgyné sz. Pólya Zsuzsanna
Pólya Mihály
Pólya Judith
Kemény Sándor
Kemény György
unokái

Lányi Antal János Mihály
dédunokája

Lakás: II. Zárda-utca 8.

Községi Temetkezési Intézet felvételi irodája VI. ker. Andr.-u. 21-23. T. 117-867.

Announcement of the death of Anna Deutsch Pólya

position for Szegő in the United States. It was clear even then that life was precarious for Jewish mathematicians in Germany and in the years that followed, mathematicians throughout much of Europe faced the same difficult decisions. The subtle and not so subtle pressures were described poignantly in a letter written by Szegő himself to Tamarkin. Szegő succeeded in leaving that year, going first to Washington University, St. Louis, and four years later to Stanford. Pólya in his 1934 letter to Tamarkin said, "The whole European situation is very dark." And it was only going to get darker. (Askey and Nevai, 1996)

Mathematical Reviews

Vol. 1, No. 1   JANUARY, 1940   Pages 1-32

ALGEBRA

Venkatarayudu, T. The 7–15 problem. Proc. Indian Acad. Sci., Sect. A. 9, 531 (1939). [MF 3]

The problem of placing $2n+1$ people around a table n times so that no two persons sit beside each other more than once is solved when $n=7$. *P. Scherk* (New York, N. Y.).

✳Dickson, Leonard Eugene. New First Course in the Theory of Equations. John Wiley and Sons, Inc., New York, 1939. ix+185 pp. $1.75.

This book retains the flavor and spirit of the author's "First Course in the Theory of Equations," but it is written in a more expansive style with many illustrations either following or preceding the introduction of new ideas or topics, and with minute attention to details.
Extract from the preface.

Sispanov, Sergio. Generalización del teorema de Laguerre. Bol. Mat. 12, 113–117 (1939). (Spanish) [MF 80]

Corliss, J. J. Upper limits to the real roots of a real algebraic equation. Amer. Math. Monthly 46, 334–338 (1939). [MF 157]

The bounds provided here are sharper but more complicated than those given by Lagrange and Jean J. Bret.
O. Szász (Cincinnati, Ohio).

Eagle, Albert. Series for all the roots of a trinomial equation. Amer. Math. Monthly 46, 422–425 (1939). [MF 204]

The author gives power-series solutions for all roots of $x^{m+n}-px^m+q=0$, m, n positive integers, p, q complex. References are given to McClintock [1895] and Birkeland [1920/21], but the important contributions of Nekrassoff [for example, Math. Ann. 29, 1887] are not mentioned. Compare also A. J. Lewis [Nat. Math. Mag. 10 (1935) (abstr. Ph.D. thesis, University of Colorado, 1932)]. The standard tool for the power-series solution of the trinomial equation, the Lagrange expansion of x in $x=h+k\phi(x)$, is used by the present author. A difficulty exists in obtaining convergent series for all roots of the equation. The various methods of solving the equations by Lagrange expansions differ in the manner in which this difficulty is overcome. The author gives what seems to be a compact and usable method. *A. J. Kempner* (Boulder, Colo.).

Eagle, Albert. Series for all the roots of the equation $(z-a)^m=k(z-b)^n$. Amer. Math. Monthly 46, 425–428 (1939). [MF 205]

As the author indicates, this equation can be transformed into a trinomial equation (let $z-a=x^n$). But it can be treated directly by reducing to a standard form $z^m=k(z-1)^n$, and applying Lagrange's expansion. In order to guarantee convergence of the series it is necessary to consider various cases according to the relations existing between m, n, k. This runs parallel to the treatment of trinomial equations [see preceding paper by the same author].

As an application, the solution of the general cubic by this method is given. *A. J. Kempner* (Boulder, Colo.).

Erdős, P. and Grünwald, T. On polynomials with only real roots. Ann. of Math. 40, 537–548 (1939). [MF 93]

Es sei $f(x)$ ein Polynom mit nur reellen Wurzeln,

$$f(-1)=f(1)=0, \quad 0<f(x)\le f(\mu) \quad \text{für } -1<x<1,$$

wobei $-1<\mu<1$, so dass μ die Stelle des Maximums von $f(x)$ im Intervall $(-1, 1)$ bedeutet. Dann ist

$$\tfrac{2}{3}\frac{2f'(1)f'(-1)}{f'(1)-f'(-1)} \le \int_{-1}^{1} f(x)dx \le \tfrac{2}{3}\cdot 2f(\mu),$$

d.h. die "Fläche der Kurve" ist enthalten zwischen 2/3 der Fläche des "Tangentialdreiecks" und 2/3 der Fläche des "Tangentialrechtecks." Das Gleichheitszeichen wird dann und nur dann erreicht, wenn $f(x)$ vom 2-ten Grad ist, d.h. nur im wohlbekannten (Archimedischen) Fall des Parabelsegments. Die Verfasser geben einen Induktionsbeweis, der nicht ganz einfach ist. Durch Weglassen des Integrals ergibt sich, dass die Höhe des Tangentialdreiecks $\le 2f(\mu)$ ist; die Verfasser gehen von diesem, von G. Szekeres gefundenen Satz und dessen einfachem Beweis aus. Man kann bemerken, dass die erste Hälfte der Doppelungleichung sich auch mit der Methode von Szekeres beweisen lässt. Es hat nämlich $f(x)$, abgesehen von einem positiven konstanten Faktor, die Gestalt

$$f(x)=(1-x^2)\Pi(\alpha_\mu-x)\Pi(\beta_\nu+x),$$

wobei $\alpha_\mu\ge 1$, $\beta_\nu\ge 1$. Man erhält, mit Benutzung der Ungleichung zwischen dem arithmetischen, dem geometrischen und dem harmonischen Mittel von zwei Zahlen, dass

$$\int_{-1}^{1} f(x)dx = \int_{-1}^{1} \tfrac{1}{2}(1-x^2)[\Pi(\alpha_\mu-x)\Pi(\beta_\nu+x)$$
$$+\Pi(\alpha_\mu+x)\Pi(\beta_\nu-x)]dx$$

$$\ge \int_{-1}^{1}(1-x^2)[\Pi(\alpha_\mu^2-x^2)\Pi(\beta_\nu^2-x^2)]^{\frac{1}{2}}dx$$

$$\ge \tfrac{4}{3}[\Pi(\alpha_\mu^2-1)\Pi(\beta_\nu^2-1)]^{\frac{1}{2}} = \tfrac{4}{3}[-f'(1)f'(-1)]^{\frac{1}{2}}$$

$$\ge \tfrac{2}{3}\frac{2f'(1)f'(-1)}{f'(1)-f'(-1)}.$$

G. Pólya (Zürich).

Collar, A. R. On the reciprocation of certain matrices. Proc. Roy. Soc. Edinburgh 59, 195–206 (1939). [MF 161]

The paper is concerned with the rapid computation of the reciprocals (inverses) of matrices of special type. Let $M=(m_{ij})$ be a moment matrix

$$m_{i+j-1}=\int_a^b w(x)x^{i+j-2}dx,$$

where $w(x)$ is a weight function. Let

$$p_i(x)=p_{i1}+p_{i2}x+\cdots+p_{ii}x^{i-1}$$

A Pólya review of a paper by Erdős on page 1 of the first issue of Mathematical Reviews *in 1940*

In the summer of 1940, Pólya received two important and welcome letters. One was dated June 11, 1940, from Szegő at Stanford where he was by then Executive Head of the Mathematics Department. Szegő wrote a letter that seems to be, by present standards, and by American customs, strangely formal, given their close earlier collaboration. On the other hand, those who knew Szegő would not be surprised at the outwardly formal tone. It is consistent with his style—he invariably wore a well-tailored suit and displayed a strikingly courtly manner. The letter reads,

My dear Professor Pólya,

In accordance with our former correspondence regarding your willingness to come to Stanford I had today a conference with President [Ray Lyman] Wilbur. He asked me to offer you a position at Stanford University as a 'research associate' for a year with a salary of $3500. This proposal is subject to the consent of the Board of Trustees which shall meet June 20. President Wilbur intends to send you the formal invitation right after the decision of the Board of Trustees. In view of the present situation in Europe he will send you the invitation by cable. Of course, this will be followed by a written statement of the invitation.

As you may remember from your former visit to Stanford, the summer quarter here extends from June 20 to September 1 and the regular school year starts September 15. Your duties which shall involve teaching and conducting of graduate work, would begin as soon as you will be able to reach Stanford University.

Hoping that you shall be able to spend the coming school year with us and looking forward to having you here, I am

very sincerely yours
G. Szegő.

This was followed on June 21 by a cable and letter of appointment along the lines described by Szegő, from President Wilbur. The appointment was for one year and that was clearly spelled out. And the rank was research associate.

This invitation was soon followed, however, by a letter from Henry L. Wriston, president of Brown University, dated July 31, 1940. This letter invited Pólya to join the Brown faculty as a lecturer for a two-year period, 1940–42. The salary was set at $2,000 annually plus $1,000 for traveling expenses for a total of $5,000 for the two years. The very temporary nature of the offer from Stanford created visa problems so, in spite of his having first accepted the Stanford offer, he subsequently had to decline and accept the offer from Brown, in order to satisfy the demands of the American consular staff in Zürich. He did acknowledge in a letter that "there [were], for the time being, serious difficulties in reaching Lisbon from Zürich," but they did get to Lisbon and went by ship, the American Export Lines's Ex Cambion, from there to New York. (Rider 1987)

Thomas Mann

Before leaving Zürich the Pólyas had been asked by the mathematician, Alfred Pringsheim, to look up his son-in-law, the novelist Thomas Mann, when they got to the United States. The Pringsheims by that time had fled from Munich and had settled in Zürich. They assumed that with the Manns in Los Angeles, this contact would be easy for the Pólyas. (This provides further evidence, if any were needed, of a common misunderstanding still prevalent in Europe—or even on the East Coast—that if one is in California, even near San Francisco, one must be just next door to Los Angeles!) As it turns out, the Manns were in New York—their permanent move to Los Angeles not scheduled to take place till March 1941.

The Pringsheims were anxious to get news to Mann that they were all right. Since they were Jewish, they had had to get out of Germany. In Munich they had been quite rich, having inherited considerable wealth. But when they fled to Zürich they could take little with them other than some paintings. This led to a clever remark by Pringsheim when he was asked how he and his wife were getting by. He answered in German, "*Von der Wand, an den Mund*" (translation: "From wall to mouth"), indicating that they were using money from the pictures that they had brought with them and sold in order to eat.

Mann had become convinced that his brother-in-law, Peter Pringsheim, the physicist, was in a concentration camp in France. (Rider, 1984) He eventually learned that this was not the case and proceeded to make what were rather complicated arrangements to have the younger Pringsheim come to the United States, where he became a research associate at the University of California, Berkeley. But in 1940 Mann was definitely concerned about his wife's family.

On arrival in New York, Pólya made arrangements to visit the celebrated writer and arrived at Mann's apartment to be greeted by Mrs. Mann. Mann then came into the room and commented after a few minutes, "You speak very good German." [1987, 2, 86] Pólya was enormously pleased, since German was not his native language, of course, and he had just been complimented by one of the most distinguished of modern German writers.

The Pólyas proceeded from New York to Providence where they spent almost two years. Toward the end of that time they spent a short time at Smith College, but this was not a particularly suitable place for Pólya since the mathematician of greatest distinction there was the ring theorist, Neil McCoy. Pólya never exhibited any genuine interest in ring theory. But while in Northampton he did some successful consulting with a local firm on the mathematics of the design of "twisting dies."

Meanwhile, Europe was engulfed in war. On March 23, 1941, Stella Pólya had a long letter from Gertrude Hardy, G. H. Hardy's sister, expressing relief that the Pólyas had gotten to America. She said, "Like you, I expected Switzerland to go. I suppose it isn't out of the soup yet." Miss Hardy went on to describe life in a "safe" part of Britain, where she was working in a school. "Of course, we have sirens most days, at first we used to go to the trenches, even at night, but now we take it in turns ... The standard of work is going down all over the country, that is inevitable because of 'black out' and all such nuisances. We have enough to eat but it is rather 'much of a muchness' and when I see an American meal scene on the screen I feel envious ... Harold has brought out 2 new books, one is delightfully readable, 'A Mathematician's Apology'. It is selling marvellously. The other is about Ramanujan ... I have no doubt it is excellent ... What about the house in Zürich, and dear little chalet at Engelberg—those were good days ... My love to you both and may we all meet again some time to do cross words. G. E. Hardy"

"Am I Really a Good Mathematician?"

While in Providence Pólya took advantage of the location to visit Norbert Wiener at MIT. Pólya shared some mathematical interests with Wiener, having worked with Plancherel, for example, on extensions of the Paley-Wiener theorem. (The original theorem relates entire functions of exponential type and Fourier transforms; the generalization involves functions of several complex variables and multiple Fourier integrals.) Earlier, in a paper of 1914, Pólya had anticipated Wiener's important Tauberian theorem. [1914, 2]

Pólya told of one such visit to Wiener, when he took the train to Boston and was picked up by Wiener at South Station. On the way back, after a day of mathematical discussions, Pólya reported that there was terrible traffic. Indicative of Wiener's well-known insecurity, Wiener stopped the car in the middle of the traffic to ask Pólya: "Am I really a good mathematician?" Pólya said later, "Have you any choice in what to answer? People were honking all around us." (Pólya 1975)

In the fall of 1942, Pólya finally arrived back at Stanford to take up an appointment as an associate professor. Prior to coming, his wife had made the long trip across the country to Palo Alto alone by train in order to locate a house for them. She found a suitable place, 2260 Dartmouth Street, within walking distance of the Mathematics Department, in what is called College Terrace, southeast of the campus in Palo Alto. They lived in this modest but charming house until Pólya's final illness in 1985. Pólya's pride was the garden, however, where he delighted in showing off his lemon and orange trees, and the loquat, which has edible flowers. While urging visitors to sample the blossoms he would recall Hilbert's famous Königsberg address of 1930 in which he cited Kronecker's remarks about number theorists being like the Lotus-eaters who, having sampled the lotus flower, can never give it up. (The Hilbert address was recorded and was made available from Springer with their *Hilbert Gedenkband* (Reidemeister 1971).)

2260 Dartmouth Street, Palo Alto

Pólya's experience in emigrating from Europe was far happier than that of many of his colleagues. His two principal appointments on arriving in the United States were both at major research universities. Such was not the fate of many emigrés who often ended up teaching large numbers of elementary classes in relatively minor institutions, at least mathematically minor in the 1930s. Max Dehn, who had solved one of Hilbert's 23 Paris problems, ended up at the Southern Branch of the University of Idaho (now Idaho State University) in Pocatello. Karl Löwner (Charles Loewner) initially found himself teaching heavy assignments of trigonometry courses at the University of Louisville. The rank of associate professor at Stanford may have been a bit of a come-down for Pólya, who had been a full professor at the ETH, but at least it was Stanford. All in all it was a lucky move.

With the arrival of Szegő the Mathematics Department at Stanford had begun to take on world-class status. Szegő brought in D. C. Spencer and A. C. Schaeffer, both mathematical luminaries, as well as Max Shiffman, Richard Bellman, and Paul Garabedian. John Herriot was appointed shortly before Pólya arrived, and somewhat later Loewner,

Szegő and Pólya on the Quad at Stanford

*Pólya's colleagues at Stanford: M. M. Schiffer, Stefan Bergman, M. H. Protter
(actually at Berkeley), John Herriot, and Charles Loewner*

Stefan Bergman, and M. M. Schiffer, attracted by the concentration on
analysis, joined the department. By that time it was highly respected,
particularly in the areas of classical analysis and applied mathematics.

The last two years before the move from Zürich, Pólya had published
rather little, a sharp contrast to the volume of work he had been turning
out in the years during and immediately after the First World War. The
time at Brown seems to have led only to the collaboration with Wiener.

With Pólya's arrival at Stanford, however, he began to develop an interest in some new problems, in some cases problems not entirely unrelated to his earlier work on conformal mapping and classical analysis.
These new problems were essentially problems of mathematical physics
and led to a series of papers, one co-authored with Szegő. This work
ultimately culminated in the monograph by Pólya and Szegő, *Isoperimetric
Inequalities in Mathematical Physics*, published in 1951 as part of the An-
nals of Mathematics Studies by Princeton University Press.

Another Drummer

In 1836 Jakob Steiner had used the symmetrization technique to prove
some isoperimetric inequalities, the simplest of which, $L^2 \geq 4\pi A$, says

that of all plane domains with given area A, the circle has the least perimeter L. Analogously, in three-dimensional space, the sphere has the least surface area for fixed volume. The planar case can be restated in the following way: If the circle has the least perimeter with fixed area, then also the circle has the maximum area for fixed perimeter. An analogous statement can be made about the sphere.

By 1899, Hermann Minkowski had developed isoperimetric inequalities between the volume and the surface area of a convex solid; between the surface area and the average width of the solid; another between the three quantities: volume, surface area, and average width. In 1856, Barré de Saint-Venant had looked at the torsional elastic prisms and observed that of all the cross sections with a given area, the circle has the maximum torsional rigidity. By 1877 Lord Rayleigh had observed that of all membranes with a given area, the circle has minimum principal frequency. And he also conjectured that of all clamped plates with a given area, the circle has the minimum principal frequency and of all conducting plates with a given area, the circle has minimum electrostatic capacity. Hadamard in 1908 also considered clamped plates and Henri Poincaré in 1903 looked for a proof that of all solids with a given volume the sphere has minimum electrostatic capacity. A number of others had looked at these problems, but the statement of the problem was advanced and much unified in a paper of Pólya and Szegő's in 1945 and ultimately by their 1951 monograph.

These questions of principal frequency have been much in the news again in the 1980s and 1990s with the publication of the striking results of Carolyn Gordon, David Webb, and Scott Wolpert, that one cannot hear the shape of a drum, that is, that one can have two differently shaped drum heads with the same principal frequency. Mark Kac had earlier written an article on this question, "Can you hear the shape of a drum?," and when he sent a reprint of his paper to Pólya, knowing that Pólya had worked on these problems, he inscribed the reprint: "From another drummer, Mark Kac." (Kac, 1966)

The 1951 monograph had a rocky start in that the material, at least in part, had been initially done as part of a research project under the auspices of the Office of Naval Research. Parts had been reported on at AMS-MAA meetings and in the *Bulletin of the American Mathematical Society* and the *American Mathematical Monthly*. A comprehensive treatment of the problem had been submitted to the American Mathematical Society but printing difficulties led to its being transferred to the newly-founded *Transactions* of the Society. Eventually the paper was withdrawn and it later took the form of the monograph described above.

Halsey Royden told of his experience with the symmetrization techniques of Pólya and Szegő at Pólya's 90th birthday dinner: "I also learned the elegant techniques and applications to analysis of the theory of symmetrization which Pólya was then developing in collaboration with Gábor Szegő. These techniques struck me as too simple and easy to be very deep until several years later at an eastern graduate school [Harvard] I was taught far more complicated methods for some of these problems which gave far more limited results than the 'elementary' methods of Pólya and Szegő."

Royden also told of his sharing bus rides with Pólya when they were both teaching classes in Berkeley, not uncommon for mathematics faculty at Stanford in the 1950s. They would board the bus in Palo Alto at 7:30 a.m. for the ride to Berkeley and for the next hour and a half Royden would be educated on topics ranging from mathematics to anecdotes about mathematicians to dissertations on the finer points of etiquette involving the familiar "du" in German and its counterpart in Hungarian. Pólya claimed that he and Szegő always spoke English in the United States because they couldn't quite sort out how to use the familiar when speaking Hungarian or German. (In later years they reverted to Hungarian.)

Speaking English did not impede progress in mathematics, however. Pólya continued to collaborate with Szegő and he did additional work in applied mathematics and analysis, often in collaboration with M. M. Schiffer, L. E. Payne, H. S. Weinberger, and I. J. Schoenberg. For details see the evaluation by Schiffer in Appendix 6.

At Stanford Pólya supervised the dissertations of eight more Ph.D. students: Madeline Johnsen (1946), Andrew Van Tuyl (1947), Burnett C. Meyer (1949), Michael Israel Aissen (1951), Grove Crawford Nooney (1953), Charles McLoud Larsen (1960), Madeleine Rose Ashton (1962), and Donald W. Grace (1964). Van Tuyl and Nooney had careers in industry. Meyer spent his career at the University of Colorado, Aissen at Fordham, Johns Hopkins and Rutgers. Sr. Madeleine Rose eventually became an administrator at the College of the Holy Names, Oakland, and Larsen became a professor of mathematics at San Jose State University. Another student of Pólya's, who took a master's at Stanford before continuing on for a Ph.D. at Notre Dame, was Lester H. Lange, who became dean of the School of Science at San Jose State. Lange remained a lifelong close friend of Pólya's, on several public occasions and in writings referring to him as "Vous sans qui mes choses ne seraient pas ce qu'elles sont" (You, without whom my affairs would not be what they are.) There are very many others who could say that as well.

High School Competitions

In 1946, obviously recalling the success of the Eötvös Competition in Hungary, Szegő and Pólya started a high school mathematics competition at Stanford. The contest continued until 1965 when the Department of Mathematics decided to shift more emphasis to graduate teaching and abandoned the competition, a decision that left Pólya somewhat embittered. The contests consisted of few, but interesting, problems and were being taken throughout the Western states with as many as 1200 participants in 151 centers. During the later years the contest was supported in part by Sylvania Electric. The problems and solutions were collected and published by Pólya and his student Jeremy Kilpatrick in 1974. Though some of the problems were later used in Pólya's problem solving books, the collection remains an impressive and entertaining one to browse through. [1974, 2]

Pólya's interests in problem solving also resulted in his being chosen, along with Tibor Radó and Irving Kaplansky, to construct the first postwar contest for the William Lowell Putnam Mathematical Competition. Pólya was a member of the Putnam Prize Committee from 1948 to 1950. In 1947, he served the Mathematical Association of America as chair of the Northern California Section, and subsequently for a three-year term, 1958–60, as a member of the MAA's Board of Governors.

Pólya's influence as a teacher extended beyond his Ph.D. students and, later, the secondary school teachers he taught in institutes. One former student, Secretary of Defense William J. Perry in the Clinton administration, took a B.S. and M.S. in mathematics at Stanford before getting a Ph.D. in mathematics from Pennsylvania State University. Pólya supervised Perry's master's thesis. In an interview in 1996, Perry said, "In college, the person who really turned me on to serious mathematics was George Pólya. He was my thesis advisor at Stanford. He was an absolutely super human being. And he took me under his wing.... I've never met—before or since—anybody quite like Pólya.... he exposed me to parts of mathematics that I had never seen before. And he was just a warm human being." (Albers 1996)

~ 10 ~

1945

How to Solve It

Alan H. Schoenfeld pointed out, in an essay that he wrote for a collection on Pólya's contributions to teaching, that 1945 was indeed a great year for problem solving. (Curcio 1987, 27–46) It was the year that Jacques Hadamard's *Essay on the Psychology of Invention in the Mathematical Field* was published by Princeton University Press. Hadamard, whom Pólya knew well and with whom he shared mathematical interests, had become interested in discovery or invention when he heard a lecture by Henri Poincaré early in the century. When Hadamard published his paper "History of Science and Psychology of Invention," he sent a copy to Pólya with the inscription: "*À G. Pólya. Hommage de haute estime et de vieille amitié. J. Hadamard*"

Karl Duncker's *On Problem Solving* also appeared in 1945. And a third work was the English translation of Max Wertheimer's *Productive Thinking*. But of all the important works on problem solving that appeared in 1945, the most influential was George Pólya's *How to Solve It: A New Aspect of Mathematical Method*.

As we have observed, of course, Pólya's interest in the techniques of problem solving began as early as 1919 and he was giving serious thought to questions of heuristics when he was working with Szegő on the *Aufgaben und Lehrsätze*. Nevertheless, *How to Solve It* was a major event in Pólya's career and in the history of heuristics as a serious subject of study.

He had written a draft of a book on the subject in German shortly before coming to America, but he later rewrote it all in English and

commented to Hardy that he was planning to call it *How to Solve It.*
Hardy said, "It is appropriate that you go to America. It is the country of
'how to' books." (Alexanderson and Lange 1987)

The reviews were favorable. The reviewer for the *Monthly* was none
other than the very well-known E. T. Bell. The price of the book was
$2.50. Bell noted that the word "new" in the subtitle was disavowed by
Pólya in the preface. Pappus, after all, had considered heuristics over
2000 years earlier. Bell says that every prospective teacher should read it,
as well as graduate students "who are required to do some teaching," and
"the traditional mathematics professor ... who might also learn something
from the book: 'He writes *a*, he says *b*, he means *c*; but it should be *d*.'"
The reissue of the book three years later—the price had gone up to $3.00—
was reviewed in *Mathematical Reviews* by no less than Hermann Weyl.

Lying as it did outside the mainstream of mathematics and of education, the manuscript did not easily attract a publisher. Pólya had to approach four before finding a willing editor, but it was finally brought out
in 1945 by Princeton University Press. For a university press, *How to
Solve It* must be one of the all-time best sellers. It has never been out of
print and many copies are still sold every year. It was brought out in a
paperback edition by Doubleday in 1957 and this enjoyed enormous
success. Later a Penguin edition appeared. For years it was something
that one might find on the racks at the drugstore or the supermarket.
Princeton has sold well over a million copies in English and there have
been translations into at least twenty-one other languages (Arabic, Bulgarian, Chinese, Dutch, French, German, Greek, Hebrew, Hungarian,
Italian, Japanese, Malaysian, Polish, Portuguese, Romanian, Russian,
Serbo-Croatian, Slovenian, Spanish, Swedish and Turkish). In a number
of these languages there have been multiple editions over the years.

Making Peace

Pólya's niece, Judi, recalls that on one of her visits to Stanford her Uncle
George took her into his study and showed her the shelf with the translations of *How to Solve It,* pointing out that "this is the way I make peace."
He had put the Hebrew and Arabic translations next to each other on
the shelf!

Pólya had an extremely high regard for Descartes' *Regulæ ad Directionem
Ingenii* (*Rules for the Direction of the Mind*). This work had been left

incomplete, but it was an attempt by Descartes to formalize the process of attacking mathematical problems. Judith Grabiner has suggested that Pólya gave too much credit to the *Regulæ* and not enough to the ideas in Descartes's more famous work *La Géométrie*. (Grabiner 1995) Unfortunately some of the later observations about *La Géométrie* were not available to Pólya. His copies of both the *Regulæ* and *La Géométrie* were, however, from their appearance, well thumbed through and used. There are also slips of paper inserted with notes. So he was certainly familiar with both works but he chose to concentrate on the *Regulæ*. That is the book he often quotes in *How to Solve It*.

Pólya himself said in *How to Solve It* that "René Descartes ... planned to give a universal method to solve problems but he left unfinished his *Rules for the Direction of the Mind*. The fragments of this treatise, found in his manuscripts and printed after his death, contain more—and more interesting—materials concerning the solution of problems than his better known work *Discours de la Méthode* [one part of which was *La Géométrie*] although the '*Discours*' was very likely written after the '*Rules*'." He goes on to say that the following words of Descartes describe the origin of the "Rules": "As a young man, when I heard about ingenious inventions, I tried to invent them by myself, even without reading the author. In doing so, I perceived, by degrees, that I was making use of certain rules." [1945, 2, 87-88]

This is not unlike some comments of Pólya's in the Preface to *How to Solve It* where he says, "The author remembers the time when he was a student himself, a somewhat ambitious student, eager to understand a little mathematics and physics. He listened to lectures, read books, tried to take in the solutions and facts presented, but there was a question that disturbed him again and again: 'Yes, the solution seems to work, it appears to be correct; but how is it possible to invent such a solution? Yes, this experiment seems to work, this appears to be a fact; but how can people discover such facts? And how could I invent or discover such things by myself?'" He went on to say that this concern led him to write the present book [*How to Solve It*], but it would also lead to his producing a more complete book on the subject, a reference, no doubt, to *Mathematics and Plausible Reasoning*, which appeared nine years later.

Pólya was certainly aware of influences on him beyond Descartes's. He was strongly influenced by Euler, as we discussed earlier, and was also familiar with the contributions of Leibniz and Bolzano in this area. In *How to Solve It* he recognized his indebtedness to some modern contributors to the subject: Ernst Mach (especially his book, *Science of Mechanics*, which for Pólya was a major inspiration), Jacques Hadamard, William James, and Wolfgang Köhler. He also mentioned the psychologist Karl Duncker. [1945, 2, 122–123]

Pólya's Rules of Discovery

Even though others were writing in this area and others had considered the questions of problem solving in earlier generations, it is nevertheless Pólya's *How to Solve It* that had the tremendous impact on the way people viewed the techniques of attacking mathematical problems. It provided an opportunity for Pólya to put interesting ideas in simple language, often with rather down-to-earth examples. Again, in the Preface, he says he challenges teachers to engage their students in the discovery of mathematics at some level. He points out that the opportunity may be lost if a student with some natural talent is not given an opportunity to try discovering mathematics and enjoying the act of discovery. He says, "He cannot know that he likes raspberry pie if he has never tasted raspberry pie. He may manage to find out, however, that a mathematics problem may be as much fun as a crossword puzzle, or that vigorous mental work may be an exercise as desirable as a fast game of tennis." Later he points out that "mathematics has two faces; it is the rigorous science of Euclid but it is also something else. Mathematics presented in the Euclidean way appears as a systematic, deductive science; but mathematics in the making appears as an experimental, inductive science."

Schoenfeld points out that *How to Solve It* and Pólya's later, more expansive books on problem solving, the two-volume *Mathematics and Plausible Reasoning* (1954) and the two-volume *Mathematical Discovery* (1962, 1964), are books that experienced people in mathematics can read with great pleasure. They look at these rules and admonitions and recognize in their own work that they have indeed used these techniques. The phrases that Pólya uses (and lists in his "short dictionary of heuristic" in *How to Solve It*) have entered the language of mathematics and mathematics education: "Could you restate the problem?"; "Did you use all the data?"; and "Draw a figure." And we are asked to consider analogy and generalization and specialization. Another suggestion is: if one cannot solve the proposed problem, can one find a problem (preferably related to the original!) that one can solve? These questions, admonitions, maxims, are in some ways common sense. Nevertheless, Pólya systematizes all of this and gives examples that make it clear how these ideas can be used to analyze and, with a bit of luck, solve problems.

Schoenfeld has demonstrated, however, that if these ideas are not specifically called to the attention of students, nothing much will happen. It is not sufficient to demonstrate the solving of problems to students, assuming that the "rules" will be assimilated automatically. If any real progress is to be made, one must point out at every turn what technique one is using on a problem—whether one is using analogy, generalization, specialization, the solution of a known but related problem, etc. In another essay on the influence of Pólya's heuristics on mathematics education, Schoenfeld analyzes the relationship between these rules and the methods used in artificial intelligence. See Appendix 7.

Clearly there are limitations in any set of rules that are to provide a method for solving problems. To the extent that the rules are general enough to apply to a wide class of problems, they may be too general to provide help in a given case; to the extent that the rules are specific, they will probably apply only to a small set of problems. The fact that Descartes gave up on his *Regulæ* and left the manuscript incomplete might indicate that he realized the limitations to devising such a set of rules. Pólya had to have realized the limitations as well. In the Preface to *Mathematical Discovery* Pólya described the praiseworthy attempts by Descartes and Leibniz to outline a universal method for solving all problems and admitted that the "quest for a universal method has no more succeeded than did the quest for the philosopher's stone which was supposed to change base metals into gold; there are great dreams that must remain dreams. Nevertheless, such unattainable ideals may influence people: nobody has attained the North Star, but many have found the right way by looking at it." [1962, 1, x] In spite of the limitations Pólya took rules for problem solving farther than anyone had to date and the fact is that his work has influenced many students, teachers, and mathematicians over the intervening years.

How to Solve It certainly became one of the great mathematical bestsellers, at least of recent times, but it is not quite clear why it has been so enormously popular, and probably considerably more widely read than his other books on problem solving, which in many ways are more interesting. The title is part of its success, promising, as it does, more than any book can reasonably deliver. The book is also short and hence inviting. The problems presented are, of course, attractive and the rules are practical and interesting. Whatever the reason, it has enjoyed enormous success.

Readers were found in unexpected places. In 1950 Lester R. Ford contacted Harold M. Bacon at Stanford about Rudolph Brandt, who was incarcerated at Alcatraz for a series of crimes but was spending his time in prison learning calculus. The prisoner had essentially no formal education. Bacon visited and corresponded with this man over a period of six years, looking over Brandt's solutions to problems and making suggestions for further study. In December 1951 Bacon took him a copy of Pólya's *How to Solve It*, which Pólya had inscribed to Brandt. The prisoner was reported to be deeply touched. (Jellison 1997)

With the appearance of *How to Solve It* and Pólya's intensifying interest in the study of method, he started at Stanford a new career, in a sense. While he continued to carry on serious mathematical research, in particular his work with Szegő on isoperimetric inequalities and joint work with various Stanford faculty members and visitors, much of his time was spent on his books on problem solving.

He officially retired in 1953 and went immediately to Switzerland where he taught some courses at the ETH. Then after the International

Robert L. San Soucie, Kenneth Ghent, Bertram Yood, Pólya, Paul Civin, and Ivan Niven on the occasion of Pólya's visit to the University of Oregon

John Hancock, Sr. Madeleine Rose Ashton, Pólya, Lester H. Lange, and Marvin Winzenread

Congress in Amsterdam in the summer of 1954 he taught for part of the year 1954–55 at Washington University, St. Louis. He lectured more and more on heuristics, notably as a visiting lecturer for the Mathematical Association of America. It was during such a visit to the University of Oregon in 1954 that I heard him lecture for the first time. My own career was much influenced by the encounter.

About this time Pólya published his popular article on picture writing [1956, 3], a beautiful introduction to generating functions. If anyone has doubts about Pólya's talent for making mathematical ideas crystal clear to a reader, that person should read this remarkable display of how to make difficult ideas easy. It is truly a *tour de force*.

Summer Institutes

In 1955 he started teaching in a series of institutes for high school and college teachers, supported by the National Science Foundation, by General Electric, and by Shell Oil. These grew out of conversations Pólya had had with Harold M. Bacon, his longtime colleague from the Department at Stanford, during a long automobile trip the two of them took

A lineup of teacher institute faculty: Pólya, Harold M. Bacon, and Cecil Holmes

together in 1953 on their way to a conference on teacher education in Boulder, Colorado, where the principal speakers were Emil Artin and Raymond L. Wilder. The first of these institutes was offered in the summer of 1955 and they continued uninterrupted until the late '60s. Most were directed by Bacon, but the Shell programs in particular were directed by a Stanford chemist, in the School of Education, Paul DeHurd. The list of visiting faculty for these summer and academic year institutes is impressive, and includes I. J. Schoenberg, Carl P. Allendoerfer, Ivan Niven, D. H. Lehmer, Morris Kline, and H. S. M. Coxeter.

One of the most interesting in the series of Stanford institutes was one held at Le Collège du Léman in Versoix, a village just outside Geneva, Switzerland. This was held in the summer of 1964 for teachers in American or international high schools in Europe, North Africa, and the Middle East. The staff consisted of Pólya, Harold Bacon, and myself. Pólya was amused that at a press conference at the beginning of the Institute, with a level of formality unfamiliar to most Americans, Bacon was consistently referred to as "Monsieur le Directeur Général" and I was "Monsieur le Directeur Adjoint," titles that stuck, in jest, for many years following the Institute. A similar institute was held the following summer, but Pólya elected to teach at Stanford instead; he was replaced by Cecil Holmes of Bowdoin College. The assistant in the programs was Pia

Pfluger, daughter of Pólya's Ph.D. student and successor at the ETH, Albert Pfluger. She later went on to become a faculty member at the University of Amsterdam, prior to which she had taught at San Diego State University and held visiting appointments at UCLA and Indiana University.

It was during the first Institute in Switzerland that Pólya observed that his successor, Albert Pfluger, had his birthday on October 13, and that the three faculty members at the Institute, myself, Pólya and Bacon had birthdays on November 13, December 13 and January 13 respectively. We formed the "13 Club" and had our picture taken together (in the correct order). Pólya was highly entertained by this coincidence.

In this long series of institutes, Pólya taught in almost all of them and was, of course, always the star attraction. It gave him an opportunity to pass along to many devoted disciples his ideas about problem solving. Through these institutes his influence on teaching was ever more widespread.

When it came to teaching, Pólya was always a showman. He used visual devices, demonstrations with concrete objects, anything that would keep the attention of his students. If he wanted to demonstrate something

The "birthday lineup": Albert Pfluger (October 13), G. L. Alexanderson (November 13), Pólya (December 13), and Harold M. Bacon (January 13)

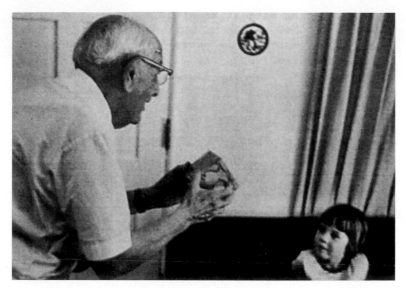

Pólya and Lisa Albers at play

with a polyhedron, he preferred to have a cardboard model rigged with rubber bands so that when the flattened model was tossed in the air, it would instantly become a 3-dimensional figure. If he was demonstrating a polygon formed with a string or rubber band, if he ran out of fingers to indicate vertices, he was not embarrassed to use his nose for the additional vertex.

Pólya was certainly an advocate of a number of approaches to teaching that are talked about a good deal today: discovery, active learning, etc. For the early grades he was an enthusiastic advocate of manipulatives — in particular, Cuisenaire rods and the geoboard. He also strongly promoted the use of the remarkable animated Nicolet films in geometry, demonstrating theorems in a way now routinely done on a computer equipped with suitable software.

Pólya on SMSG

In the immediate post-Sputnik era Pólya had been an outspoken critic of the formalism of the School Mathematics Study Group (SMSG) and the "new math." He often cited an example taken from the SMSG geometry text that gave a theorem with proof—taking up half a page—stating that with three points on a line, one point must lie between the other two. He argued that though this is a necessary theorem for a foundations course in geometry, it has no place in an elementary text. He said that had he been asked to study the proof of such a theorem in high school, he would almost certainly have given up on mathematics, having concluded that the subject is dumb! In support of this viewpoint—that there was excessive rigor in the "new math"—he was one of the signers of a manifesto on curriculum reform that appeared in *The Mathematics Teacher* and the *American Mathematical Monthly* decrying the direction of the reform movement of the 50's and 60's.

In talking about what he regarded as the excess of rigor in SMSG, he cited the oft-repeated story about Isadora Duncan's proposal of something like marriage to George Bernard Shaw. She argued that their children would have his intelligence and her beauty. Of course, Shaw pointed out that the children might well have.... Pólya suggested that SMSG had been put together by research mathematicians and high school teachers, on the assumption that the material would reflect the mathematical sophistication of the researchers, and the pedagogical skills of the high school teachers. But then, like Shaw, he pointed out that the material in fact reflected.... The observation was unkind but there was, perhaps, some truth in it.

Pólya was equally distressed, if not more so, by the "back to basics" backlash that came after the "new math." This was followed by the problem solving approach, something he would certainly have approved of (albeit in a somewhat different form, since some educational materials only paid lip service to "problem solving," used as a buzzword to promote material that was often somewhat thin in content).

Basically, Pólya was a humanist. He saw doing mathematics and solving problems as a human activity, creative and exciting. To teach his ideas, he referred back to earlier masters, often quoting the wisdom of philosophers and scientists of the past. Among his papers we find lists of quotations, listed in an order that appears not to be accidental. Here's a sample:

1. In the discovery of lemmas the best aid is a mental aptitude for it.   —Proclus

2. Examples are better than precepts for learning the arts.
                                                                                —Newton

3. All this is done naturally, and is sometimes better done by those who never learnt the precepts of logic than by those who learnt them.   —Logique de Port-Royal

4. Thus I stick to the idea that a plain man having the advantage of these helpful precepts and of some practice could surpass the very best, just as a child can draw better lines with a ruler than the greatest master without such artificial aid.   —Leibniz

5. Reasoning, as a practical process, must be learnt by practice, in the same manner as any other practical art, for example riding, or fencing. We are not secured from committing fallacies by such a classification of fallacies as logic supplies, as a rider would not be secured from falls by a classification of them.   —Whewell

6. The helps offered to improve the mind consist in certain ways of thinking which facilitate thinking.   —Leibniz

7. At each step that we take it should be comprehensible to the reader why we take that step.   —Bernard Bolzano

8. Yet the most perfect and cogent foundation of an idea is reached when all the ways and the motives which led to it and consolidated it are clearly set forth.   —Ernst Mach

9. It is of great advantage to the student of any subject to read the original memoirs on that subject, for science is always most completely assimilated when it is in the nascent state.
    —J. C. Maxwell

10. He (Euler) preferred the instruction of his pupils to the little satisfaction of amazing them. He would have thought not to do enough for science if he should have failed to add to the discoveries with which he enriched science the candid exposition of the ideas which led him to those discoveries.   —Condorcet

And others were:

… increasingly convinced that it is not knowledge, but the means of gaining knowledge, which I have to teach.   —Dr. Arnold

Good logic can be the worst enemy of good teaching.
    —Cf. Descartes, Feynman, Minsky

The ideas should be born in the student's mind and the teacher should act only as midwife.   —Socrates

And elsewhere he wrote down a quote of Condorcet, in discussing Euler:

Mathematics is the science that yields the best opportunity to observe the working of the mind. Its study is the best training of our abilities as it develops both the power and the precision of our thinking. Mathematics is valuable on account of the number and variety of its applications. And it is equally valuable in another respect: By cultivating it, we may acquire the habit of a method of reasoning which can be applied afterwards to this study of any subject and can guide us in life's great and little problems.

> —From "Éloge de M. Euler par le Marquis de Condorcet." See Leonhard Euler, *Opera*, ser. 3, vol. 12, p. 287.]

The quotes from Condorcet were two of Pólya's all-time favorites and he used them often.

Of course, Pólya's advice was always to come back to "guessing and proving." When he signed his name with his initials, G. P., he would sometimes add in parentheses "Guessing and Proving", another manifestation of Pólya's irrepressible impishness, always present in his conversation and in his letters.

A caricature of Pólya by William J. Firey, 1952.

~ II ~

Honors

Recognition came late to Pólya, but it did come. Oddly enough, probably the greatest honor that he would be given came first, on the seventh of July, 1947. It was on that day that the Académie des Sciences, Paris, of the Institut de France, elected him to membership as a corresponding member in the geometry section, to replace Charles de la Vallée Poussin, a Belgian, who at that time was made an *"associé étranger."* (De la Vallée Poussin and Hadamard had independently proved the Prime Number Theorem in 1896.) Few Americans were chosen by the Académie; even at the time of Pólya's death there were only 36 Americans in the Académie, and only eight of those were mathematicians.

That same year Pólya received his first honorary doctorate, a Doctor of Mathematics degree, from the Eidgenössische Technische Hochschule, Zürich. Further honorary degrees were a Doctor of Laws from the University of Alberta in 1961; an honorary Doctor of Science from the University of Wisconsin, Milwaukee, in 1969; and an honorary Doctor of Mathematics degree from the University of Waterloo in 1971.

Additional memberships in national academies came considerably later than that of the Paris Academy: the Académie Internationale de Philosophie des Sciences, Bruxelles, in 1965; the American Academy of Arts and Sciences, 1974; the National Academy of Sciences (U.S.), 1976; and an honorary membership in the Hungarian Academy of Sciences, also in 1976. Many felt that the membership in the National Academy of Sciences (U.S.) should have come much earlier. In fact, Paul J. Cohen, when he was elected to the Academy, felt that it was a scandal that Pólya

INSTITUT   DE FRANCE

ACADÉMIE   DES SCIENCES

Paris, le 7 juillet *19* 47

Les Secrétaires perpétuels de l'Académie des sciences,
à Monsieur George POLYA, Correspondant de l'Académie des
sciences de l'Institut de France.

Monsieur,

Nous avons l'honneur de vous informer que l'Académie
des sciences de l'Institut de France vous a élu, en la
séance de ce jour, correspondant pour la section de géométrie en remplacement de M. Charles de LA VALLEE POUSSIN,
élu associé étranger.

Veuillez agréer, Monsieur, avec nos cordiales féli-
citations, l'assurance de nos sentiments de haute estime.

Announcement from the Académie des Sciences, Paris

was not a member and proceeded to work for his election. Though membership in the Paris Academy came when Pólya was 60, he was 89 by the
time the National Academy got around to electing him. The reason for
this, in part, could have been the oft-expressed concern in earlier years
that someone's chances were better for membership in the National Academy if one were at an Eastern university. Now that seems to have changed.

Announcement from the American Academy of Arts and Sciences

Nevertheless, there was a sense of outrage over the National Academy's long delay in recognizing Pólya's work. Stanislaw Ulam wrote, "I do not need to tell you how I was glad and feeling relieved from the embarrassment of being a member, after your election, which I always felt was at least half a century overdue." Other members wrote: "[We] congratulate the National Academy for electing you a member," "I felt a mixture of pleasure and shame when I heard about your election to the National

22 Dec 1956

KING'S COLLEGE CAMBRIDGE

Dear Professor Pólya,

I have great pleasure, as President of the London Mathematical Society, in informing you that you were elected an Honorary Member of the Society at the Annual General Meeting.

With the highest regards,

Yours Sincerely

P. Hall

Letter from Philip Hall

Academy of Sciences. It is absolutely incredible that the Academy could have waited so long. What is wrong with the system? Finally they have corrected this terrible mistake for not having you voted in 30 or 40 years ago. Better late than never, but I still cannot help feeling deeply ashamed for this fact," "It came as a great shock to me to learn that you had just been elected to the National Academy, since I had taken it for granted that you had been a member for a long time. I am very happy that this wrong has now been righted." There were other letters in a similar vein, from members of the Academy and others. Pólya's response to his well-wishers was: "I would not say that my election to the National Academy came too early. Yet your kind words, and some other similar friendly letters are a great compensation. Many, many thanks!"

In addition, in 1952 Pólya was made an honorary member of the Council of the Société Mathématique de France; an honorary member of the London Mathematical Society in 1956; an honorary member of the Swiss Mathematical Society in 1957; an honorary life member of the Mathematical Association (of Great Britain) in 1972; an honorary member of the Society for Industrial and Applied Mathematics that same year; an honorary life member of the New York Academy of Sciences in 1976; and an honorary life member of the California Mathematics Council, also in 1976. The written notifications of some of these awards are interesting. For example, the notification from the Académie des Sciences, Paris, has the obligatory engraved portrait of Marianne, representing France, while the American Academy of Arts and Sciences is very "early American" in appearance with lots of engraved Spencerian script. It is signed by the Secretary after the phrase "Respectfully your obedient servant." The appearance of the announcement is suitable for an organization that was founded in 1780 "under the leadership of John Adams." Some of the notifications are elaborate certificates. By contrast, the notification of honorary membership in the London Mathematical Society is in the form of a small handwritten note from Philip Hall on the letterhead of King's College, Cambridge.

"I Like the Company"

In 1963 the Mathematical Association of America gave Pólya its second Award for Distinguished Service to Mathematics at its annual meetings in Berkeley. The first such award had gone to Mina Rees the year before. This is the MAA's most prestigious award. That same year Stanford renamed the building then housing the Department of Computer Science; they called it Pólya Hall. (Before that it had been called Hemlock Hall!)

In 1965 Stefan Bergman attended the World's Fair in New York and found in the IBM pavilion a demonstration of random walk. (This exhibit—or a copy—was on display later at various science museums—Seattle, Los Angeles, among others.) Prominently displayed were the names of the seven principal contributors to random walk: Albert Einstein, Enrico Fermi, Norbert Wiener, Andrei Kolmogorov, John von Neumann, George Pólya, and Stanslaw Ulam. Pólya remarked, "I like the company I am in." [1987, 2: 155]

There were other honors along the way as well. In 1968 his film, "Let Us Teach Guessing", won the Blue Ribbon Award (top honor) in the

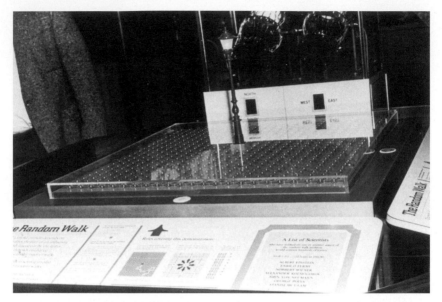

The Random Walk Exhibit at the IBM Pavilion, New York World's Fair

Mathematics and Physics category at the Educational Film Library Association's First Annual Film Festival in New York. This film, distributed then as now by the Mathematical Association of America, showed Pólya in a classroom discussing before a class the partition of space problem of Steiner's.

For Pólya's 60th birthday celebration in 1947, there was an elaborate proclamation produced with extensive calligraphy and bound in red leather, referring to Pólya's many accomplishments up to that time. The document was signed by 134 well-wishers. The list includes many Stanford colleagues, including the physicist and Nobel laureate, Felix Bloch, A. H. Bowker, John Herriot, Charles Loewner, Halsey Royden, Don Spencer, Charles Stein, Gábor Szegő, and Frederick Terman, but also an extraordinary list of mathematical luminaries from around the world: H. F. Bohnenblust, Richard Courant, Harold Davenport, Griffith C. Evans, Willy Feller, Maurice Fréchet, J. R. Kline, Emma and D. H. Lehmer, Richard von Mises, Otto Neugebauer, John von Neumann, Jerzy Neyman, Hans Rademacher, Raphael and Julia Robinson, Arthur Rosenthal, I. J. Schoenberg, J. J. Stoker, Ottó Szász, Oswald Veblen, Abraham Wald, J. L. Walsh, Hermann Weyl, Norbert Wiener, and Antoni Zygmund, among many others. It was quite an event.

Gábor Szegő and George Pólya with Mrs. Szegő (left) and Mrs. Pólya (right)

For Pólya's 75th birthday, seven of his colleagues produced a Festschrift entitled, *Studies in Mathematical Analysis and Related Topics,* published by the Stanford University Press. The editors were his colleagues Gábor Szegő, Charles Loewner, Stefan Bergman, M. M. Schiffer, Jerzy Neyman, David Gilbarg and Herbert Solomon. It contains contributions from 60 distinguished mathematicians.[1]

In 1976, the Mathematical Association of America (MAA) established its George Pólya Awards for expository writing in the *College Mathematics Journal.* The list of distinguished recipients includes Anneli Lax, G. D. Chakerian, Douglas R. Hofstadter, Paul R. Halmos, Philip J. Davis, Constance Reid, Paulo Ribenboim, and Pólya's former student, Lester H. Lange, among many others. A few years earlier, in 1971, the Society for Industrial and Applied Mathematics (SIAM) had established a Pólya Award in Combinatorics and Its Applications, prompted by contributions to the Society by Frank Harary and William T. Tutte. The effort to establish the prize was led by Harary. This award, accompanied by a bronze medal designed by Mrs. Donald E. Knuth, has been awarded to a

---

[1] Looking over the responses to the letter of invitation sent by Szegő to potential contributors to the Festschrift, one is struck by the formality of most of the salutations. Only eleven used "Dear Gábor", while eight used "Dear Szegő", a convention surely declining in usage when addressing a colleague. Sixteen played it safe with "Dear Professor Szegő", one used "Mr. Szegő". And then there were the more exotic greetings: "Lieber Herr Szegő", "Cher collègue", "Cher monsieur", "Sehr geehrter Herr Kollege", and "Kedves Professzor Ur."

SIAM Pólya Medal (obverse and reverse)

number of outstanding combinatorialists. The list of winners includes some impressive names, including Ronald L. Graham, Richard P. Stanley, Endre Szemerédi, László Lovász, and Sarahon Shelah. (The Hungarians are well represented!) In 1994, the breadth of the award was expanded to include in alternate years specialists in other fields. (See Appendix 9.)

Two years after Pólya's death, the London Mathematical Society established its Pólya Prize awarded for the first time in 1987, "to an individual in recognition of outstanding creativity in, of imaginative exposition of, and of distinguished contribution to, mathematics within the United Kingdom." This too has now been given to some distinguished mathematicians: the first two awards went to John Horton Conway and C. T. C. Wall. The Prize is given in years when the Society is not awarding its most prestigious prize: the De Morgan Medal.

In 1990, the Mathematical Association of America named one of its two main headquarters buildings at 1527 Eighteenth Street, N.W., in Washington, DC, the Pólya Building. Participants in the dedication ceremony included Pólya's nephew, Anthony Lányi, the son of Suzanne Lányi and grandson of Pólya's brother Jenő. The plaque on the Pólya Building in Washington reads:

> "Whereas, George Pólya distinguished himself as a mathematician par excellence in real and complex analysis, probability, combinatorics, geometry, number theory, and mathematical physics; and
>
> "Whereas, Professor Pólya had a long and distinguished career as teacher, educator, and mentor to thousands of students; and

*Pólya in his study at home
on Dartmouth Street*

"Whereas, he contributed in many ways to the advancement of the Mathematical Association of America; and

"Whereas, he greatly enhanced mathematical knowledge through his numerous articles and books, especially in the area of mathematical discovery; and

"Whereas, Professor Pólya is admired and loved by his colleagues in the mathematical community;

"Therefore, I, Lida K. Barrett, President of the Mathematical Association of America, and acting on behalf of the Board of Governors of the Association, do dedicate this property as the George Pólya Building as part of the Dolciani Mathematical Center in this city of Washington, DC, the nation's capital."

At the present time, the building is occupied in part by offices of the MAA and in part by offices of other mathematical organizations, including the Joint Policy Board for Mathematics, the Conference Board of the Mathematical Sciences, and the Washington offices of the American Mathematical Society.

In 1990, the MAA also set up a Pólya Lecturership to send distinguished mathematical expositors to meetings of the twenty-nine Sections of the

MAA, visiting each section roughly every five years. Again those desig-
nated as Pólya Lecturers in the first six years of the program set a high
standard: John Ewing, Patricia K. Rogers, Carl Pomerance, Robert
Osserman, Underwood Dudley and László Babai. (See Appendix IX.)

Pólya has also been recognized by a number of groups of teachers, as
well as by major educational organizations. In 1972, he was honored at
the Second International Congress on Mathematical Education (ICME)
in Exeter, England. Eight years later, when the Fourth ICME was held in
Berkeley, California, he was named honorary president. And in 1988,
three years after his death, he was recognized at the ICME in Budapest,
where Jean-Pierre Kahane gave a talk summarizing Pólya's contributions
to mathematics and teaching.

Of course the lasting tribute to a mathematician-writer is his legacy of
ideas. Jorge Luis Borges put it this way: "When writers die, they become
books, which is, after all, not too bad an incarnation."

~ 12 ~

The Later Years

Pólya retired in 1953, having held the rank of full professor at Stanford for only seven years. Some thought that his promotion to full professor came too slowly. It is reported that Jerzy Neyman at Berkeley prompted the promotion by threatening to violate the "gentlemen's agreement" between Berkeley and Stanford that they not raid each other's faculties. Pólya, during the Second World War, had commuted to Berkeley to teach Neyman's probability courses while he was away doing work associated with the war effort. Pólya thus got to know the Berkeley faculty reasonably well and they got to know him. With the threat that he might be recruited to Berkeley, Stanford promoted him to full professor in 1946. (Reid 1982, 211)

Retirement did not imply much of a change in Pólya's schedule, however. He continued to do research and also came back to teach various seminars and, very often, classes. This required action by the Provost of the University, since the rule up to that time specified that an emeritus professor could not continue to teach. But Pólya prompted the change of policy and proceeded to teach till he reached the age of 90!

Stanford's longtime provost, Frederick E. Terman, often credited with the growth of "Silicon Valley," said of the new Stanford policy that allowed faculty to stay on after age 65 and foreshadowed revised federal retirement legislation: "The best example . . . was [George] Pólya, who at his prime was the senior mathematician on the West Coast, and who was an absolutely superb teacher at intermediate level, at advance level, and even for teachers of high school math. The fact is we kept Pólya teaching until he was over 80 and he earned every dime he received." (Glover 1979)

Following Pólya

In the fall of 1955, when I entered Stanford as a graduate student, Pólya was teaching the problem seminar for first-year students, in cooperation with Loewner. From time to time he also taught courses in combinatorics. George Dantzig tells that he was once scheduled to teach combinatorics, but then it was arranged that Pólya would teach the first half of the course, Dantzig the second. Dantzig claims that the enrollment went from 13 to 50 to 120 when word got out that Pólya would be teaching. On the first day of class, when Pólya started to lecture, one of the students leaned over to Dantzig, who was sitting in on the class, and said, "I'd certainly hate to be the guy who follows Pólya."

Pólya's last course in combinatorics was taught jointly with Robert E. Tarjan, again with Pólya teaching the first half of the course, the material up through the Pólya enumeration theorem, appropriately enough. Pólya was ninety years old at the time. The notes for the course, written up by Donald R. Woods, were of sufficient interest that they were published as a successful separate monograph by Birkhäuser in their Progress in Computer Science series. [1983, 1]

A. Wayne Roberts, in reviewing the book said that "among first rank research mathematicians ... very few manage to establish beyond the borders of their own schools a reputation for teaching. George Pólya did." He went on to say that these notes by Woods "are different (and better) than the usual notes in which a well-intentioned scribe of the Gaussian school has carefully removed the scaffolding before exposing to us an austere collection of definitions, theorems, and proofs. These notes preserve the dynamics of the classroom, the motivational techniques, even the sense of timing that lets us in on what questions were posed just before a class ended.... And we see that Pólya sees classroom opportunities to enunciate such problem solving maxims as 'if you cannot solve the proposed problem, solve first a suitable related problem.' ... One is struck throughout the notes with Pólya's willingness to risk being didactic, to put into words the general principles that we perhaps too frequently suppose students will notice without our stating them in so many words.... How would our teaching be affected if, as class proceeded, we kept a running list on the side board of each principle illustrated by our activities in front of the class that day?" (Roberts 1986) This indicates that Pólya indeed did spell out in detail the problem solving techniques being used, the procedure that Alan Schoenfeld has shown to be effective in his studies of problem solving in teaching.

Books Are Like Children

Partial retirement did mean that Pólya had more time to work on books. Shortly after retirement he published the two-volume set, *Mathematics and Plausible Reasoning*. In 1962 and 1964, the two volumes of *Mathematical Discovery* appeared, but between these two volumes he produced the first version of *Mathematical Methods in Science* as Volume XI in the Studies in Mathematics series by the School Mathematics Studies Group (SMSG). It was edited by Professor Leon Bowden of the University of Victoria (Canada) and was based on a course that Pólya had developed for teachers in the institute program at Stanford. This book was reissued in 1977 in the New Mathematical Library series of the Mathematical Association of America. The pace at which he produced these manuscripts contradicted his cautionary remarks about taking on writing projects. He had often quoted Beaumarchais: "*Les livres sont comme les enfants des femmes, conçus avec volupté, menés-à-terme avec fatigue, enfantés avec douleur.*" (Books are like children — conceived with pleasure, carried with weariness, born with pain.)

Pólya's influence in the area of heuristics extended beyond his own books and articles, however. He suggested to Imre Lakatos that he write his 1961 dissertation for Cambridge on the Euler-Descartes formula for polyhedra. The formula states that if V is the number of vertices, E the number of edges, and F the number of faces of a polyhedron, then $V - E + F = 2$, though, as it turns out, only when certain conditions are put on the polyhedron itself. Lakatos later turned this into a series of articles in the *British Journal for the Philosophy of Science*, and after his death it was published as a separate book by the Cambridge University Press: *Proofs and Refutations/The Logic of Mathematical Discovery*. The book is truly a *tour de force*, in the form of a very complex Socratic dialogue. It's an impressive display of erudition (the footnotes alone are worth a look) and it conveys something of the process of doing mathematics, the interplay between proving and refining the statement of what one set out to prove.

By 1974, Pólya and a former Stanford colleague, then at the University of Virginia, Gordon Latta, produced a textbook in complex analysis, *Complex Variables*. The book has had many admirers, but Pólya was never completely satisfied with it. In my copy his inscription included a well-known quote of Martial from the *Epigrammata*: *"Sunt bona, sunt quaedam mediocria, sunt mala plura. Quae legis hic: aliter non fit, Avite, liber."* (There are good things, there are some indifferent, there are more things bad that you read here. Not otherwise, Avitus, is a book produced.)

The book did have an impact, however, in the teaching of complex analysis. A simple device, a special vector field for giving a geometrical and physical interpretation to complex integrals, was introduced. This was picked up and expanded on by Bart Braden in 1983 and 1987, for one of which Braden won the Allendoerfer Award. (Braden 1987) These Pólya vector fields are especially suited for visualizing complex integrals and particularly helpful now, since they can be portrayed by easily available computer graphics packages. A number of applications appear in Needham's book on visualizing complex analysis. (Needham 1997, 481–505)

But always there were interesting problems. On August 24, 1974 he wrote me from Flims (Graubünden) a letter with the following combinatorial problem: "There are $4n$ players who wish to play bridge at n tables. Each player must have another player as partner, and each pair of partners must have another pair as opponents. Show that the choice of partners and opponents can be made in exactly $(4n)!/n!8^n$ different ways. (3, 315,... ways if $n = 1, 2,...$—always an odd number)". At 87 he was still inventing problems.

One advantage to retirement was that Pólya's schedule was more flexible and he and Stella could travel and spend long periods in Europe. When they went back to Zürich they always stayed at the Kurhaus Zürichberg, a long tradition. They continued to have good friends there. In particular, Stella Pólya was close to Hurwitz's daughter Lisi and they were also close to the widow of one of Pólya's Ph.D. students, Alfred Aeppli. And they spent time with Albert and Maria Pfluger, and Joseph Hersch.

They always traveled to Europe by ship from New York or Montreal—keeping souvenirs of trips on Cunard's R. M. S. Mauretania and R. M. S. Queen Mary, as well as the American Export Lines's S. S. Constitution—and continued this tradition till 1964. These trips also involved rail trips across the country with stops along the way in Cleveland to consult on the Greater Cleveland Mathematics Project, with a visit to his niece, Susi Lányi, in nearby Oberlin, Ohio, and in New York for consulting at the

George Pólya, Susi Lányi, Stella Pólya at Oberlin, ca. 1965

Rockefeller Institute (now Rockefeller University). They retained contact with friends in the Mathematics Department at NYU, later to become the Courant Institute. This department was in many ways the successor to the great institute at Göttingen and Pólya felt a strong mathematical attachment to the group in New York. Of course, two of the leaders at NYU were J. J. Stoker, Pólya's former Ph.D. student in Zürich, and, later, Peter Lax, a nephew of Szegő's.

The first time Pólya flew in an airplane was in 1964 when the Pólyas were in Zürich and an invitation came to speak at a conference in England. There was insufficient time to get to the conference by surface transportation, so he agreed to fly—and he found that he actually enjoyed it. After that the Pólyas flew between San Francisco and Zürich. The title of his talk in England was, incidentally, "How I Did Not Prove the Riemann Hypothesis"!

Courant

About Richard Courant himself, Pólya had some reservations. Anneli Lax, a student of Courant's, had told Constance Reid: "In his case one is quite aware, you know, that he's not perfect. He does some things that are really quite objectionable, and yet you like the man. Perhaps it is because in spite of the somewhat devious ways the kinds of things he really stands for and accomplishes are O.K. I mean, they are the things you would want to do too, although you would do them straightforwardly. I am not a person who believes that any means is justified by the end, not at all, but he seems to get away with it. Somehow." (Reid 1976, 290) Pólya shared these views, and even though he and Courant had known each other for many years, they were not close. Pólya did tell, though, about the time when they both attended a conference in the Colorado Rockies in the early 1950s and, while there, did some climbing. Pólya told Courant about a Swiss tradition, that when climbing, upon reaching a certain altitude, people would use the familiar form of address in German, "*du.*" Of course, it was understood that when they descended they would return to using the more formal "*Sie.*" Courant misunderstood and thought Pólya was proposing that henceforth they use "*du*" with each other. Pólya had meant no such thing. But Courant seemed pleased and Pólya decided to leave it at that.

Davenport

In 1958, the International Congress of Mathematicians took place in Edinburgh. The Pólyas went to England after a summer lecturing in Vancouver and were met by Harold and Anne Davenport who drove them through Scotland on an extensive automobile trip prior to the Congress. This turned out to be a high point in their lives and they often recalled what a wonderfully pleasurable occasion it was. C. A. Rogers wrote that Davenport had spent the year 1947–48 at Stanford, and though he says that Davenport had "acquired a taste for the American way of life, and made lifelong friends of Pólya and of Szegő, it is scarcely possible to detect any influence of this visit on his work." (Rogers 1971) This may be an accurate statement, but it does ignore the fact that at that time Pólya and Davenport did write a joint paper on products of series [1949, 1], not an entirely inconsequential result of the visit. (I like the paper: for years it gave me an Erdős number—the length of a chain of coauthorships to Paul Erdős—of three, subsequently reduced to two!) At any rate, the Pólyas were enormously fond of the Davenports. After Pólya's death, Stella sent an extraordinary Mappin & Webb silver fruit bowl to Anne Davenport, a bowl she had bought almost 30 years earlier on their automobile trip through Scotland.

In 1983 Anne Davenport, recalling the good times they had had when she and her husband visited Stanford, wrote in a letter prompted by a look at a preview of the *Pólya Picture Album* that had been published in the *College Mathematics Journal*: "I relive so many happy times—the period when I first came into contact with mathematicians—we had the experience of being entertained by [Alfred] Errera in the 'little' house left to him after the war. (The grand piano was just an insignificant trifle in their 'small' drawing room!) When they came to England I was so nervous that I forgot to put any salt in the vegetables! Harald Bohr visited us in London. In France we met Hadamard with his story about the period of the Dreyfus affair when Hermite sent for him and said *'Vous êtes un traître, Monsieur Hadamard.'* [You are a traitor, Monsieur Hadamard.] Poor Hadamard was very nervous till Hermite went on *'vous avez quitté la théorie des nombres* [you have abandoned number theory]....' Norbert Wiener with his long stories. But then I turned the page and got to your American stage and at once I was back in that truly wonderful year in Stanford—one of the happiest of my life." The story of Hadamard takes on added poignancy when one realizes that Dreyfus's wife was Hadamard's cousin.

Harold and Anne Davenport

Budapest—54 Years Later

Because of the problem that Pólya had perceived over the years resulting from his not responding to the call for military service in the First
World War, he had never been back to Budapest since that time. (His fear
of going back earlier was probably not justified, since the Communist
regime would probably have approved of his reluctance to fight for the
Hapsburgs!) Anyway, in 1967, for the first time in 54 years, he went to
Budapest and Stella accompanied him. He remarked, upon returning,
that he was enormously pleased that, though he had spoken little Hungarian, except occasionally with Szegő, in the intervening years, when
he arrived in Budapest his Hungarian came back to him very quickly, if
indeed it had ever left. Pólya was very good at languages, speaking not
only his native Hungarian, but German (both high German and the Swiss
dialect), French, English, and some Italian. Because his wife was from the
French-speaking part of Switzerland, they always spoke French at home
and there are indications that French was really his favorite language. His
papers were written in Hungarian, French, German, English, Italian, and,
curiously, in one case, in Danish. Of course, in addition to the languages
just mentioned, he was reasonably accomplished in Latin and classical Greek.

*Jenő, John, and
George Pólya,*

Jenő Pólya with his wife Lili

For many years Pólya had not used the accent on the "o" when sign-ing his name, but in his later years at Stanford he went back to using it, making his name appear more obviously Hungarian. He never attempted to explain it, even when his wife chided him about it. She thought it appeared nationalistic and silly. And she told him so several times in my presence. But he only smiled.

Returning to Budapest must have been, in some ways, difficult. There were few members of his family left. His brother Jenő had been shot by the fascists late in the war, sometime during December 1944 or January 1945. His sister Ilona had died in 1938, Flóra in 1950. He was to return to Hun-gary once more, in 1977, and, though he planned to visit again to attend a conference in 1980, health problems intervened and he was unable to go.

During the 1950s and 60s there was a regular flow of European math-ematicians visiting the Stanford Mathematics Department. For lectures, there were Richard Courant, J. E. Littlewood, Alfréd Rényi, Wacław Sierpiński and Jan Van de Corput, among others.

Stefan Bergman was not considered a good teacher and taught very little at Stanford, but he prided himself on his linguistic skills, and I recall two visits by Sierpiński when he lectured in French. Bergman generously volunteered to translate each time, so a sentence or two of Sierpiński's would be followed with an extended (and no doubt embellished) transla-tion into English by Bergman. Unfortunately, is was easier to understand Sierpiński's French than it was to understand Bergman's English.

I vividly recall the visit by Littlewood because it was viewed by faculty and graduate students alike as something of an event. I was a third-year

Pólya and J. E. Littlewood at the time of Littlewood's 1957 visit to Stanford

graduate student at that point and recall going to the lecture and finding every seat taken in the lecture hall, so I had to stand at the back. The front row was occupied by senior professors, including Pólya. It was a hot day and classrooms at Stanford were not air conditioned, at least in the old buildings on the Inner Quad. Littlewood was wearing tweeds appropriate for Cambridge, perhaps even for the Hebrides! So he looked quite miserable and proceeded to shed his jacket and his tie, even unbuttoning his shirt. And then for something well over an hour he rapidly covered the board with mathematics. I understood absolutely nothing of what he said. The next day I encountered Pólya on the Quad and, in an effort to make small talk, I commented that it was an impressive thing to see Littlewood but I understood nothing of his proof. Pólya said reassuringly, "Don't worry about it. I never figured out what he was trying to prove." It gave me courage to know that even longtime co-authors do not necessarily understand each other's colloquium talks.

About this time, motivated by some data collected by D. H. and Emma Lehmer, Pólya wrote a paper [1959, 3] on the similarity of reasoning in number theory and physics. The problem he examined is one of finding

Pólya, Julia Robinson, Raphael Robinson, Stella Pólya

the probability, P_d, that positive integers x and $x + d$, where d is an even integer, are both primes. The simplest case, of course, where $d = 2$, concerns twin primes. Pólya was able to give a heuristic argument that

$$P_d = 2C_2 \prod_{p|d} \frac{p-1}{p-2} \frac{1}{(\log x)^2},$$

where $C_2 = \prod \left(1 - \frac{1}{(p-1)^2} \right)$, a formula conjectured by Hardy and Littlewood in 1922 and still unproved.

I. J. Schoenberg

Also in 1957–58, I. J. Schoenberg came to Stanford to spend the year and he and I shared Pólya's office in one of the mathematics buildings on the Inner Quad. The office, though large, was rather gloomy and showed no evidence of the remodeler's hand since the building was built in the early 1890s. The most interesting thing about it was a set of small framed pictures of all the universities where Pólya had studied or held appointments: Budapest, Vienna, Göttingen, Institut Henri Poincaré, the ETH, Trinity College, Cambridge, Fine Hall at Princeton, Brown, and Stanford. I always found that impressive. But for work Pólya preferred his study at home. But more of that later.

Pólya had known Schoenberg since 1930. Schoenberg's first wife was Dolli, the daughter of Landau, thereby adding evidence of the oft-cited traditional hereditary principle in mathematics that mathematical talent is passed from father to son-in-law. (Pólya would probably modify that now to say that mathematical talent is passed along from mother to daughter-in-law.) Karlin has pointed out that Schoenberg in his early career had been a dutiful son-in-law and had devoted a number of his papers to solving problems that had been initiated by Landau. Later he became best known for his work on splines in numerical analysis. But during his time at Stanford he worked with Pólya on some problems in analysis based on a paper of Pólya's on probability. [1930, 2] Schoenberg developed a whole series of papers on what he referred to as Pólya frequency functions. These are, as Karlin points out, "totally positive kernels of the form $K(x, y) = f(x - y)$ where x, y traverse the real line (or the integers)." (Karlin 1973) Schoenberg in a series of six papers explored the whole theory of these frequency functions and used, in the process, a number of results of Pólya's, as well as theorems by Laguerre and Schur. As a result of this work in 1958, Pólya and Schoenberg produced a joint paper on cyclic Pólya frequency functions, that is, such functions defined on a circle. It turns out there are numerous applications of these in analysis. It was in a letter of March 17, 1947, that Schoenberg informed Pólya that he had named these functions the Pólya frequency functions. He also pointed out in the letter that the functions occur as limiting distributions in statistics.

Though Pólya and Schoenberg collaborated on only one paper, the friendship was a long and warm one. The Pólyas were very fond of Schoenberg and of his wife, Dolli. (Schoenberg's second wife, a concert pianist, was also named Dolly, spelled differently from the name of his first wife, who had died early on.)

"Old One, Don't Complain"

Pólya's study, alluded to earlier, was a small but extraordinarily comfortable and visibly mathematical sort of room. He had framed engravings of heroes on the walls: Descartes, Euler, Lagrange, Laplace, Pascal, and Leibniz. There were also framed pictures of special favorites, some contemporary, on the walls and on the desk: the Bolyais, father and son,

Stella Pólya, Helen Crozier, G. L. Alexanderson, Pólya

G. H. Hardy, Issai Schur, and, of course, Adolf Hurwitz. On the floor to ceiling bookshelves there were many volumes from Euler's *Opera Omnia*, particular volumes from the collected works of Leibniz and Descartes, among others. The rest of the house was less obviously mathematical, but was clearly the home of cultivated and interesting people: Persian carpets, a good record collection, lots of books (many on art but many of literature no longer very fashionable: Anatole France, Victor Hugo, among others more obscure) and on the walls of the hallway a long series of framed colored woodcuts of the Minnesingers. Also, in the hallway there was a little sign reading, "*Alte, brumme nicht* [Old one, don't complain.]."

Travel destinations during these years included not only central Europe. In September, 1968 the Pólyas went off to the Caribbean where he was one of the principal speakers at a Commonwealth conference on mathematics education held at the University of the West Indies in Trinidad. Both were impressed by the exotic vegetation and the animals, especially the birds. And in 1972, Pólya was honored, along with Jean Piaget, at the Second International Congress on Mathematical Education in Exeter. The Congress proceedings reported that when the organizing committee tried to come up with principal speakers, there were two names that surfaced over and over: Pólya and Piaget. "The term 'heuristic'—the study of the methods and rules of discovery and invention—is for many inseparably linked with [Pólya's] name," the phrase "concept formation" with Piaget's. (Howson 1973, 9) (It came as a happy

Gábor Szegő

surprise to discover at the Congress that Stella Pólya had been a class-
mate of Piaget's in dancing class when they were children in Neuchâtel!)

Let Us Teach Guessing

In 1968 Pólya had been awarded a prize for his film "Let Us Teach Guessing". While in England for the Exeter Congress, he made a film "Guessing and Proving," produced for the Open University by the BBC. The
first film dealt with the Steiner problem of finding the maximum number of (finite and infinite) 3-dimensional cells formed by cutting 3-space
by five planes. For the second film he considered proving the Pythagorean
theorem in a "Lecture without Words," consciously using Mendelsohn's
language, but instead he finally chose to consider the question of finding the polyhedron with the largest volume determined by n points on a
sphere. It's a problem not unlike the isoperimetric problem for polyhedra. It was shown by Michael Goldberg in 1935 that among the regular
polyhedra three are the "best" isoperimetrically — the regular tetrahedron,
the hexahedron and the dodecahedron. The other two, the regular octahedron and icosahedron, are not the "best," i.e. they do not contain

maximum volume for fixed surface area. What the "best" look like is not known. (Goldberg 1935) The problem examined by Pólya in the film was this: with 4, 6, 8, 12 or 20 points on the sphere and the convex hull formed, in which cases does the regular polyhedron formed contain the most volume? Here the regular tetrahedron, octahedron and icosahedron are the "best," and the others are not. The results of the two problems are, in a sense, dual and are certainly counterintuitive and surprising. For eight points the double hexagonal pyramid is "better" than the hexahedron, for example, and Pólya demonstrated this. But it is not the "best." That was found by two of Victor Klee's students in 1970. (Berman and Hanes 1970)

In the 1970s travel was becoming more difficult for the two of them. Still, in 1972, Pólya reported on a trip they had taken to Rome where they relived seeing all the sights in one of their favorite cities and where Pólya had to give a lecture in French ("but repeating some words and phrases in Italian") to a group of teachers. While there he also made another film on teaching, prompted by his friend Bruno de Finetti. In the film Pólya had to carry on a conversation with a puppet named Giorgio—in Italian!

The next year, while in Zürich, he came down with a particularly bad case of flu and had to spend 11 days in a "nice Swiss hospital." He was slow in recovering but was pleased that his doctor there described him as "old man in pretty good health." He was 86 at the time.

Though he was slowing down physically, he was determined to remain active to the extent possible, with as little help as possible. He pointed out that in Prussia, while it was de rigeur to help a colonel put on his great coat, to help a general would have been a grave faux pas. It could imply that he was too old to cope unaided.

In 1974, the first two volumes of Pólya's *Collected Papers* appeared; the third and fourth volumes were not to appear until 1984. All were published by the MIT Press. The editors had the same problems that anyone writing about Pólya will have. He worked in so many branches of mathematics, it is difficult, if not impossible, for a single editor—or the author of a biography, as this example will confirm—to be acquainted with all of these fields. In the case of the *Collected Papers* the solution was to find editors for the various volumes and then parcel out papers to specialists for commentary. In the case of Volume I, entitled *Singularities of Analytic Functions*, the editor was Ralph P. Boas, Jr. The commentary on the papers in the volume was done by Boas and E. Hille, L. Ahlfors, M. S. Robertson, and H. Wittich. That was a fairly easy one. Volume II, *Location of Zeros*, was again edited by Boas and the commentaries on the papers were done by M. Marden, I. J. Schoenberg, and M. Kac. Then things started to get complicated and this probably explains, at least in part, the delay of ten years in the appearance of the third and fourth volumes. The third volume, labeled simply *Analysis,* was edited jointly by Joseph Hersch and Gian-Carlo Rota. Here the papers spanned a much wider range as one sees from the list of those who provide the commentary: R. P. Boas, G. Strang, I. Richards, S. Chowla, G. E. Andrews, R. Brauer, D. Willett, E. Hille, H. Ziegler, I. J. Schoenberg, J. Hersch, P. Lax, R. C. Read, A. C. Zaanen, and E. M. Wright. Volume IV was equally complicated in that it spanned probability, combinatorics, and teaching and learning in mathematics. There was a single editor in this case, Gian-Carlo Rota, and commentary was provided by L. Santaló, H. S. M. Coxeter, J. F. C. Kingman, S. Karlin, R. M. Dudley, F. Spitzer, H. E. Kyberg, Jr., D. Klarner, S. G. Williamson, E. M. Palmer, R. C. Read, and F. Harary.

The work of the editors in compiling these collected papers is truly impressive. The task was, as indicated, daunting because of the enormous spread. Also, in providing commentaries, the editors hoped to indicate what the influence of the various papers had been and this requires a deep knowledge of the subject, whence the various specialists consulted.

As with many collected papers there are omissions, and in a few cases it can appear mysterious as to why certain papers are not included. Eight of Pólya's earliest papers are in Hungarian and these are not included in the *Collected Papers,* but some of these appeared later, in somewhat revised form, in German or French. Nevertheless one paper in Hungarian, that seems not to have been reprinted and was not included in the *Collected Papers,* did at least stimulate some interest. The paper that he wrote in 1914 on positive quadratic forms with a Hankel matrix [1914, 7] prompted work by Szász. (Szász 1921)

In 1974 the *Aufgaben und Lehrsätze* was recognized on its fiftieth anniversary. At the national meetings of the American Mathematical Society and the Mathematical Association of America that took place that year in San Francisco, Springer gave a special luncheon to recognize the event. Prior to that, of course, Pólya and Szegő had already been hard at work preparing the English translation, the first volume of which appeared in 1972, the second in 1976. This translation contained a number of new problems. At the luncheon one of the speakers was Donald E. Knuth, who had noted the upcoming anniversary and had prompted the organization of the luncheon. He referred to the *Aufgaben und Lehrsätze* as "one of the most influential books ever written about higher mathematics."

George Pólya explaining something to the author, Santa Clara, 1960s

That same year there were Stanford ceremonies recognizing Pólya's election to two national academies and in San Francisco the Hungarian Consul made the presentation of the certificate for membership in the Hungarian Academy. Pólya returned to his early interest in probability to write a delightful little note [1976, 1] on two proofreaders who read the same manuscript. One detects A misprints, the other detects B misprints, and both detect C misprints. He calculates an estimate of the number of misprints that remain unnoticed at the end of the process. It was his penultimate mathematical paper (though he did later write other pieces on problem solving and education).

"Never Say a > b..."

In 1977, to celebrate Pólya's 90th birthday, the Mathematics Department at Stanford arranged quite a grand dinner at the Stanford Faculty Club with a number of distinguished speakers: Nobel laureate Felix Bloch, a former student; Jerzy Neyman; Halsey Royden; and Donald Knuth. Peter Lax, at that time President-elect of the American Mathematical Society, read a letter of congratulations. He recalled that for many years in Hungary Pólya was "proudly known as the talented kid brother of the

Pólya in a classroom in the 1960s; Jean Pedersen on the right

famous surgeon Eugene Pólya. I would like to relate a few matters I learned from you as your student in a reading course in the summer of '46 which sheds some light on your mathematical attitudes and personality. You dismissed my suggestion to read through 'Inequalities,' by Hardy, Littlewood and Pólya saying that it is a tool, not a subject, and turned instead to the theory of entire functions. You stressed the importance once of bringing out the algebraic aspects of a mathematical subject, even when the subject is analysis. I remember much good advice in a lighter vein, such as to respect the alphabet and never say $a > b$, if you can help it."

Knuth pointed out that 90 is not a very interesting number. Powers of 2 are interesting but no one had recognized in any special way Pólya's 64th birthday. But on August 31, 1977, Pólya had been exactly 2^{15} days old! (Knuth, 1977)

A few years later, in 1980, Pólya was named honorary chair of the Fourth International Congress on Mathematical Education when it met in Berkeley. He was to give the opening address, but was unable to do so. I read his opening remarks for him at the Congress because shortly before the meetings he experienced an attack of shingles over the top of his head and down toward one eye that caused him a great deal of pain and curtailed all activities and travel. After a time it reduced his eyesight to the point where he could no longer read without the help of a device involving a television set that enlarged images on a screen so that he could read letters and articles. He maintained an interest in keeping up on things, but he had to do it under difficult circumstances. His condition continued to deteriorate, though gradually, but he still very much enjoyed the visitors, who came frequently. One of the most faithful was Professor Jean Pedersen of Santa Clara University, who stopped by regularly to discuss mathematics with Pólya and to visit both Pólya and his wife. For some years Pedersen had worked with Pólya on some questions involving polyhedra, the Euler characteristic and Descartes's angular deficiency. The Schiffers were also regular visitors as were many others in the Stanford department. Out of town visitors included nieces and his nephew, John, the Pflugers from Zürich and the ever-traveling Paul Erdős.

Finally in the summer of 1985 Pólya suffered a stroke. The situation became more and more difficult. His niece Susi Lányi came from Ohio and his niece by marriage, Kati Pataki, came from Budapest to help out. During the first few weeks of his final illness he was able to carry on somewhat confused conversations, one of the most consistent concerning ideas for proving the Riemann hypothesis. He spent some time at Stanford Hospital, eventually being transferred to a convalescent hospital, Lytton Gardens in Palo Alto, where he died on September 7, 1985, one month to the day after the death of Szegő. Following cremation the ashes were taken to Alta Mesa Cemetery in Palo Alto. Stella Pólya died on February 27, 1989, also at Lytton Gardens.

Epilogue

A few years before Pólya's death the venerable apricot tree in the garden had died. Pólya went out and bought a young shoot to replace it, acknowledging that it takes an optimist to plant, in his mid-nineties, a new fruit tree. But Pólya was always an optimist. Erdős gave an account of his conversation with Pólya when he called to congratulate him on his 97th birthday: "I told him you will celebrate your hundredth birthday with great splendor." Pólya answered: "Maybe I want to be a hundred, but not a hundred one because old age and stupidity are very unpleasant." (Schechter 1998, 196)

Paul Erdős

A memorial service was held on October 30, 1985, at the Memorial Church, Stanford. Pólya would not have cared very much about a memorial service and his widow was ambivalent about it, but she finally decided to go along with the idea, since, she concluded, the Stanford community would expect something. The service was primarily musical, with a program of Beethoven (G minor Violin and Piano Sonata, Op. 96) and Schubert (A major Quintet, Op. 114), with a moving tribute by Max Schiffer. The formal University Memorial Resolution, written by Harold Bacon, Solomon Feferman, John G. Herriot, Halsey Royden and myself, was published on December 11 of that year.

In May of the following year meetings in Pólya's honor were scheduled in quick succession. On May 3, 1986, the California Mathematics Council–Northern California, held meetings honoring Pólya at Santa Clara University and the following weekend Stanford had a memorial symposium. These back-to-back conferences made clear Pólya's dual role in mathematics—the second concentrating on Pólya's research contributions, the first on his influence on teaching. There were many speakers at the first conference but a sampling, with the titles of their talks, gives something of the flavor of the presentation: Henry L. Alder (U.C. Davis), "Pólya's Legacy: Problems and Competition, Some Reminiscences with Illustrations"; Nicholas Branca (San Diego State), "Teaching Guessing/Pólya's Style"; Larry Hawkinson (Gunn High School), "What I Learned from Pólya: Induction, Then Mathematical Induction"; C. M. Larsen (San Jose State), "Area Ratios in Convex Polygons as Pólya-Led Voyage of Geometric Discovery"; John Mitchem (San Jose State), "Pólya's Counting Theorem"; and Jean Pedersen (Santa Clara), "Geometry in the Larger Sense, Combinatorics".

The Stanford symposium lasted two days and featured Albert Baernstein II (Washington, St. Louis), Kai Lai Chung (Stanford), Albert Edrei (Syracuse), Paul Erdős (Hungarian Academy of Sciences), Ronald L. Graham (AT&T Bell Labs), Dennis Hejhal (Minnesota), and Alan H. Schoenfeld (U.C. Berkeley), with a guided tour of Pólya's photograph album by myself. At the dinner following the Symposium, Erdős spoke and in a letter to Stella Pólya, written the day after the dinner as Erdős was flying to Atlanta, he expanded on some of his remarks: "In my short speech after dinner … I said that I first met Pólya in 1934 on the way from Hungary to England. I stopped off in Zürich to see him. We met next in Oslo in 1936 and then frequently in the U.S. He met me at the station in Palo Alto when I first arrived there in December 1945 (perhaps you remember on new years day in 1946 I had lunch at your house.

We had goose and home grown guavas.) I attended a talk of Pólya about 10 years ago in Hungary where he started the talk by saying 'I am too old to tell you new theorems; I will just talk about mathematics in general' and he went on and gave a very nice talk. I told to a friend soon I will be old too and will not talk about new theorems and will give a talk like Pólya. He said laughing: 'Do you hope to give a talk as good as Pólya?' I ended by quoting from 1001 Nights: In the stories there they usually greeted the king by saying, 'King, may you live forever.' More realistically a Mathematician should be greeted by saying, 'Mathematician, may your theorems live forever.' I am sure Pólya's theorems will live forever.... Kind regards, au revoir, Paul Erdős." Erdős was an extraordinarily kind man and this comes through in this letter.

On April 8–11, 1987, the National Council of Teachers of Mathematics held its annual meeting in Anaheim and dedicated it to Pólya. The meetings were titled "Learning, Teaching, and Learning Teaching," the title of a 1963 article of Pólya's in the *American Mathematical Monthly*. Featured speakers included Lester H. Lange on "Pólya's 'Mushrooms' and Rules of Teaching and Style. Max-min Results without Calculus," Jeremy Kilpatrick on "Is Teaching Teachable? George Pólya's Views on the Training of Mathematics Teachers," and myself on "George Pólya—Mathematician and Teacher." The theme throughout was problem solving.

In August 1988 the International Congress on Mathematical Education was held in Budapest and, appropriately, it was dedicated to Pólya. A medal with Pólya's likeness was struck and distributed at the Congress. The President of the International Commission on Mathematical Instruction,

ICME Budapest medal (obverse and reverse)

Jean-Pierre Kahane, of the University of Paris-Sud, devoted his opening address ("La Grande Figure de Georges Pólya") to a detailed description of Pólya's work in mathematics and in pedagogy. He spoke at length of Pólya's contributions to the art of mathematical discovery, but has added something about Pólya discussed little elsewhere—his interest in politics, in elections and voting procedures. He further noted Pólya's pacifist period and the fact that shortly before his death Pólya had joined in an international appeal for a nuclear armaments freeze. (Kahane 1989)

Stella Pólya stayed on at 2260 Dartmouth Street with the help of live-in companions—Margaret (Gretl) Brown, Kati Fekete and Nelly Domingo—and the help of friends like Jean Pedersen, Annie Uhrbach, and Tibor Scitovsky. And Paul Erdős continued to visit on his regular trips to Stanford. Stella always commented on how kind he was. Adjustment was not easy after 67 years of marriage, but Stella Pólya was a remarkable woman, intelligent, cultivated and strong. During the Second World War she had taught French at Stanford because there was need, but except for that period, her career, consistent with her time and her upbringing, had been to support her husband's career, which she did with fierce loyalty and pride. Politically she was active in the Women's International League for Peace and Freedom. In Zürich she had studied bookbinding and a number of books in the Pólyas' library were bound in handsome bindings of linen and boards, in the Swiss style. And in 1977 she published her first (and only) article. It appeared in *California MathematiCs*, a pedagogical journal of the California Mathematics Council, affiliate of the National Council of Teachers of Mathematics. It was called "Pin Cushion Geometry for Primary Children" and concerned the placing of pins in various patterns in a pin cushion with associated activities to involve children. (Taylor and Taylor 1993, 133)

One can only guess what kind of career she might have had had she been born later and into a less conventional family. In the last few years of her life, even with failing eyesight, she wanted to keep up with what was going on in the world of ideas. When a friend went to Paris, she asked her to send her books from France that would give some idea of what was happening in French literature and philosophy.

Pólya's musical taste was conservative: symphonic and chamber music from the Romantic school—Beethoven, Chopin, Schubert primarily—and the operas of Puccini. Through all their years of marriage Stella listened dutifully to this safe music, knowing essentially nothing of music after Wagner. A year or so after Pólya's death she watched on television a particularly violent German production of Strauss's *Elektra*. She was enormously impressed by the power of it — "thrilled" is not too strong a word to describe her reaction. For weeks she wanted to discuss it. I found it startling and a bit sad that someone was discovering for the first time in the 1980's a successful and popular opera written in 1909.

To the extent her health permitted she looked out for her husband's legacy. She made contributions to the Société Mathématique de France and the London Mathematical Society and she participated in the planning of a large exhibit on Pólya in the Bender Room of the Stanford University Library, organized by the Department of Special Collections to recognize the deposit of Pólya's papers in the Stanford Archives. Stella attended the opening of the exhibit, which followed a talk by Robin Rider, sponsored by the Stanford Historical Society, on "The Whirlpool of Events: The Immigration of Scientists to Stanford and Berkeley, 1933–1945." Several friends spoke at the opening in the Library: Harold M. Bacon, Lester H. Lange, and M. M. Schiffer. It was held on what would have been Pólya's one-hundredth birthday, December 13, 1987. Dean Lange quoted appropriately from Saroyan's *The Time of Your Life*: "In the time of your life, live … so that in that good time there shall be no ugliness or death for yourself or for any life your life touches.… In the time of your life, live—so that in that wondrous time you shall not add to the misery and sorrow of the world, but shall smile to the infinite delight and mystery of it."

Donald E. Knuth, after seeing the exhibit, "remarked that he considered it the best he had ever seen on a mathematician." (Stanford Historical Society 1987) A small part of this exhibit remains on display at the Stanford Mathematics Library, along with a portrait of Pólya in oils by Milwaukee artist Elsie C. Hervis.

Many of Pólya's admirers and disciples would agree with the homage paid him in 1977 by Frank Harary who wrote: "With no hesitation, George Pólya is my personal hero as a mathematician … [he] is not only a distinguished gentleman but a most kind and gentle man: his ebullient enthusiasm, the twinkle in his eye, his tremendous curiosity, his generosity with his time, his spry energetic walk, his warm genuine friendliness, his welcoming visitors into his home and showing them his pictures of great mathematicians he has known—these are all components of his happy personality. As a mathematician, his depth, speed, brilliance, versatility, power and universality are awe inspiring. Would that there were a way of teaching and learning these traits." (Harary 1977)

Poster for the Pólya exhibit at the Stanford University Library

The most touching tribute was that of Jeremy Kilpatrick, a former student and coauthor, who wrote shortly after Pólya's death: "If you had been a graduate student living in College Terrace in Palo Alto, you might have

Stella and George Pólya

had a chance to spend every summer afternoon in the shaded study of the modest house on Dartmouth ..., drinking fruit juice and eating apples and learning lessons about being a scholar—the need to avoid self-pity and self-importance; to take pains with your work; to find the right word, the right idea; to see the fun, the humor in what you are doing. And when you returned in later years with your family and saw Pólya amusing your five-year old with a folding toy or picking lemons for you in the yard, you might have caught a glimpse of the truth that great teachers do not simply teach us to do; they teach us to be." And, he wrote further, "... a master teacher does not follow the syllabus as much as invent it; he does not cover the ground, he makes it bloom." (Kilpatrick 1985)

References

Aigner, Martin, and Günter M. Ziegler, *Proofs from THE BOOK.* Berlin, Springer, 1998.

Albers, Donald J., "The mathematician who became Secretary of Defense," *Math Horizons,* September, 1996, pp. 14–21.

Albers, Donald J., and G. L. Alexanderson, *Mathematical People, Profiles and Interviews.* Boston, Birkhäuser, 1985.

Alexanderson, Gerald L., *The Pólya Picture Album/Encounters of a Mathematician.* Boston, Birkhäuser, 1987.

Alexanderson, Gerald L., and Leonard F. Klosinski, *A History of the Northern California Section/Mathematical Association of America/1939–1988.* [Santa Clara, CA], 1988.

Alexanderson, Gerald L., and Lester H. Lange, "Obituary/George Pólya," *Bull. London Math. Soc.* 19 (1987), 559–608.

Alexanderson, Gerald L., and Kenneth Seydel, "Kürschák's tile," *Math. Gaz.* 26 (1978), 220–25.

Alexanderson, Gerald, et al., "George Pólya 1887–1985," *Campus Report,* December 11, 1985, p. 22.

Allen, Thomas Tracy, "Pólya's orchard problem," *Amer. Math. Monthly* 93 (1986), 98–104.

Askey, Richard, and Paul Nevai, "Gábor Szegő: 1895–1985," *Math. Intelligencer* 18 (1996), 10–22.

Bekes, Robert A., "An infinite version of the Pólya Enumeration Theorem," *Inter. J. of Math. Math. Sci.* 15 (1992), no. 4, 697–700.

Berge, Claude, *Principles of Combinatorics.* New York, Academic Press, 1971.

Berman, Joel D., and Kit Hanes, "Volumes of polyhedra inscribed in the unit sphere in E," *Math. Ann.* 188 (1970), 78–84.

Biggs, H. L., E. K. Lloyd, and R. J. Wilson, *Graph Theory/1736–1936.* Oxford, Oxford University Press, 1976.

Boas, Ralph P., Jr., "Selected topics from Pólya's work in complex analysis," *Math. Mag.* 60 (1987), 271–74.

——, "George Pólya/1887–1985," *Biographical Memoirs* (National Academy of Sciences) 49 (1990), 338–55.

Braden, Bart, "Picturing functions of a complex variable," *College Math. J.* 16 (1985), 63–72.

——, "Pólya's geometric picture of complex contour integrals," *Math. Mag.* 60 (1987), 321–27.

Burkill, Harry, "G. H. Hardy (1877–1947)," *Math. Spectrum* 28 (2) (1995–96), 25–31.

Burkill, J. C., "Albert Edward Ingham, 1900–1967," *Biographical Memoirs of Fellows of the Royal Society* 14 (1968), 271–86.

Carlson, F., "Über Potenzreihen mit ganzzahligen Koeffizienten," *Math. Z.* 9 (1921), 1–13.

Cartwright, Dame Mary, "Some Hardy-Littlewood manuscripts," *Bull. London Math. Soc.* 13 (1981), 273–300.

Chowla, S., Letter to G. Pólya, December 13, 1931.

Császár, Ákos, "A mathematical competition of a hundred years," *Mathematical Competitions* 10 (1997), 69–73.

Curcio, Francis, ed., *Teaching and Learning/A Problem-Solving Focus.* Reston, VA, National Council of Teachers of Mathematics, 1987.

Davenport, Harold, "A. E. Ingham," *Nature* 216 (1967), 311.

de Bruijn, N. G., "Omzien in Bewondering," *Nieuw Arch. Wisk.* (4) 3 (1985), 105–19.

Dent, Bob, *Hungary.* New York, W. W. Norton, 1990.

Erdős, Paul, Letter to Mrs. Stella Pólya, May 11, 1986.

Farkas, József, ed., *Mindenki Ujakra Készül*, volume 4. Budapest, Magyar Tudományos Akadémia, 1967.

Feller, William, *An Introduction to Probability Theory and Its Applications,* 3rd edition. New York, John Wiley and Sons, 1968.

Fischer, Gerd, et al., *Ein Jahrhundert Mathematik 1890–1990. Festschrift zum Jubiläum der DMV.* Braunschweig/Wiesbaden, Deutsche Mathematiker-Vereinigung, 1990.

Glover, Frederick O., "Fred Terman's paper trail; a gold mine of scientific research," *Stanford Observer,* April 1979, p. 5.

Goldberg, Michael, "The isoperimetric problem for polyhedra," *Tôhoku Math. J.* 40 (1935), 226–36.

Grabiner, Judith, "Descartes and Problem-Solving," *Math. Mag.* 68 (1995), 83-97.

Guggenbühl, Gottfried, ed., *Eidgenössische Technische Hochschule/1855–1955/École Polytechnique Fédérale.* Zürich, Buchverlag der Neuen Zürcher Zeitung, 1955.

Harary, Frank, "Homage to George Pólya," *J. Graph Theory* 1 (1977), 289–90.

Hardy, Gertrude Edith, Letter to Stella Pólya, March 23, 1941.

Hardy, Godfrey Harold, *Collected Papers of G. H. Hardy.* Oxford, Clarendon Press, 1966–79.

——, "A letter of G. H. Hardy," *Math. Mag.* 63 (1990), 312–13.

Haselgrove, C. B., "A disproof of a conjecture of Pólya," *Mathematika* 5 (1958), 141–45.

Hecke, Erich, Letter to Ludwig Bieberbach, January 27, 1921.

Hersh, Reuben, and Vera John-Steiner, "A visit to Hungarian mathematics," *Math. Intelligencer* 15 (2) (1993), 13–26.

Honsberger, Ross, *Mathematical Gems/From Elementary Combinatorics, Number Theory, and Geometry.* Washington, Mathematical Association of America, 1973.

Howson, A. G., ed., *Developments in Mathematical Education. Proceedings of the Second International Congress on Mathematical Education.* Cambridge, University Press, 1973.

Jellison, Charles, "The prisoner and the professor," *Stanford Magazine*, March-April 1997, pp. 64–71.

Kac, Marc, "Can you hear the shape of a drum?" *Amer. Math. Monthly* 73 (1966), 1–23.

Kahane, Jean-Pierre, "L'Adresse du Président: La grande figure de Georges Pólya," *Proceedings of the International Congress on Mathematical Education* (Ann and Keith Hirst, ed.), 1989, pp. 81–97.

——, "George Pólya 1887–1985," Supplement 1986, *Encyclopædia Universalis*, Paris, Encyclopædia Universalis, 1987.

Kanigel, Robert, *The Man Who Knew Infinity: A Life of the Genius Ramanujan.* New York, Scribner's, 1991.

Karlin, Samuel, "To I. J. Schoenberg and his mathematics," *J. Approx. Theory* 8 (1973), vi–ix.

Kemeny, John G., "How to teach guessing," *Review of Metaphysics* 9 (1956), 638–42.

Kilpatrick, Jeremy, "An editorial," *J. Res. Math. Ed.* 16 (1985), 323.

——, "George Pólya's influence on mathematics education," *Math. Mag.* 60 (1987), 299–300.

Knuth, Donald E., "George Pólya, the Hungarian nonagenarian," unpublished manuscript, 1977.

——, "Introductory remarks by D. E. Knuth at the Pólya-Szegő luncheon. Thursday, January 17, 1974," unpublished manuscript, 1974.

Kolata, Gina, "Remembering a 'Magical Genius'," *Science* 236 (1987), 1519–21.

Korevaar, Jaap, "John Edensor Littlewood (9 June 1885–6 September 1977)," *Jaarboek van de Koninklijke Nederlandse Akademie van Wetenschappen* 1977, pp. 1–5.

Kürschák, József, *Hungarian Problem Books I and II, based on the Eötvös Competitions, 1894–1905*. Washington, Mathematical Association of America, 1963.

Lakatos, Imre, *Proofs and Refutations: The Logic of Mathematical Discovery*. Cambridge, Cambridge University Press, 1976.

Lax, Peter D., "The differentiability of Pólya's function," *Adv. in Math.* 10 (1973), 456–64.

Lehman, R. Sherman, "On Liouville's function," *Math. Comp.* 14 (1960), 311–20.

Lukacs, John, *Budapest 1900: A Historical Portrait of a City and Its Culture.* New York, Weidenfeld and Nicholson, 1988.

Macrae, Norman, *John von Neumann.* New York, Pantheon, 1992.

Maz'ya, Vladimir, and Tatyana Shaposhnikova, *Jacques Hadamard, The Universal Mathematician.* Providence, American Mathematical Society, 1998.

McCagg, William O., Jr., *Jewish Nobles and Geniuses in Modern Hungary.* New York, Columbia University Press, 1972.

Needham, Tristan, *Visual Complex Analysis.* Oxford, Oxford University Press, 1997.

Neyman, Jerzy, "A glance at the history of mathematical statistics in the U.S.", unpublished transcript of speech at the IMS-WNAR meeting at Stanford University, June 23, 1977 (A).

——, "My encounters with Pólya", unpublished manuscript, 1977 (B).

Niggli, Paul, "Die Flächsymmetrien homogener Diskontinuen," *Z. Krystall.* 60 (1924), 283–98.

O'Brien, Katharine, *Excavation and other verse.* Portland, ME, Anthoensen Press, 1967.

Pekonen, Osmo, "Unkarin matematiikan uusi Eukleides ja uusi Arkhimedes," in *Kulttuurin Unkari.* Jyväskylä, Atena Kustannus Oy, 1991.

Pfluger, Albert, "George Pólya," *J. Graph Theory* 1 (1977), 291–94.

Pinl, Maximilian, "Kollegen in einer dunklen Zeit. III.," *Jber. Deutsch. Math.-Verein.* 73 (1971/72), 153–208.

Plancherel, Michel, "Mathématiques et mathématiciens en Suisse (1850–1950)," *Enseign. Math.* 6 (1960), 194–218.

Pólya, George, Letter to Ludwig Bieberbach, January 4, 1921.

——, Letter to S. Chowla, January 5, 1932.

——, Transcript of conversation (unpublished), November 5, 1975.

——, Transcript of interview (unpublished), February 26, 1978.

Pólya, John Béla, Letter to G. L. Alexanderson, July 28, 1986.

——, Letter to G. L. Alexanderson, July 14, 1988.

Read, R. C., "Pólya's theorem and its progeny," *Math. Mag.* 60 (1987), 275–82.

Redfield, J.H., "The theory of group reduced distributions," *Amer. J. Math.* 49 (1927), 433–55.

Reid, Constance, *Courant: Göttingen and New York/The Story of an Improbable Mathematician.* New York, Springer-Verlag, 1976.

——, *Neyman—from Life.* New York, Springer-Verlag, 1982.

Reidemeister, Kurt, *Hilbert/Gedenkband.* Berlin, Springer, 1971.

Ribenboim, Paulo, "Prime number records," *College Math. J.* 25 (1994) 280–90.

Rider, Robin, "Alarm and opportunity: Emigration of mathematicians and physicists to Britain and the United States, 1933–1945," *Historical Studies in the Physical Sciences* 15 (1984), 107–176.

——, Transcript of address delivered at Stanford University, December 13, 1987.

Roberts, A. Wayne, Review of *Notes on Introductory Combinatorics,* by George Pólya, Robert E. Tarjan, and Donald R. Woods. *Amer. Math. Monthly* 93 (1986), 494–96.

Rogers, C. A., "Harold Davenport/1907–1969," *Biographical Memoirs of Fellows of the Royal Society* 17 (1971), 159–92.

Rowland, Ingrid, "Titian: The Sacred and Profane," *The New York Review of Books,* March 18, 1999.

Royden, Halsey L., "A history of mathematics at Stanford," *A Century of Mathematics in America, Part II,* (Peter Duren, ed.) Providence, RI, American Mathematical Society, 1989.

——, Remarks on the occasion of Pólya's 90th birthday, unpublished, 1977.

Sagan, Hans, *Space-Filling Curves.* New York, Springer-Verlag, 1994.

Sándor, J., and I. M. Modlin, "Eugene Pólya: The Billroth of Budapest," *J. Amer. College of Surgeons* 181 (1995), 352–62.

Schattschneider, Doris, "The Pólya-Escher connection," *Math. Mag.* 60 (1987), 293–98.

——, *Visions of Symmetry.* New York, W. H. Freeman, 1990.

Schechter, Bruce, *My Brain Is Open: The Mathematical Journals of Paul Erdōs.* New York, Simon and Schuster, 1998.

Schiffer, Menahem Max, "George Pólya (1887–1985)," *Math. Mag.* 60 (1987), 268–70.

Schoenfeld, Alan H., "Pólya, problem solving, and education," *Math. Mag.* 60 (1987), 283–91.

Stanford Historical Society, "Exhibit, reception honor Pólya," *Sandstone and Tile* 12 (1) (1987), 23.

Szász, Ottó, "Über Hermitesche Formen mit rekurrierender Determinante und über rationale Polynome," *Math. Z.* 11 (1921), 24–57.

Szegő, Gábor, "Award for Distinguished Service to George Pólya," *Amer. Math. Monthly* 70 (1963), 1–2.

——, *Collected Papers.* (Richard Askey, ed.) Boston, Birkhäuser, 1982.

Szegő, Gábor, et al., *Studies in Mathematical Analysis and Related Topics*. Stanford, Stanford University Press, 1962.

Szy, Tibor, ed., *Hungarians in America*. New York, Kossuth Foundation, 1966.

Tanaka, Minoru, "A numerical investigation on cumulative sum of the Liouville function," *Tokyo J. Math.* 3 (1980), 187–89.

Taylor, Harold, and Loretta Taylor, *George Pólya/Master of Discovery*. Palo Alto, CA, Dale Seymour Publications, 1993.

Titchmarsh, E. C., "Godfrey Harold Hardy/1877–1947," *Obituary Notices of Fellows of The Royal Society* 6 (1949), 447–61.

Ulam, S., "John von Neumann 1903–1957," *Bull. Amer. Math. Soc.* 64 (1958), 1–49.

Weyl, Hermann, Letter to Ludwig Bieberbach, January 16, 1921.

Wieschenberg, Agnes, "Remembering George Pólya (1887–1985)," *Mathematics in College*, Spring-Summer 1986, 6–9.

——, "A conversation with George Pólya," *Math. Mag.* 60 (1987), 265–68.

Pólya's Bibliography

* Denotes book or monograph.

The number in parentheses at the end of an entry is the number assigned to the work in the *Collected Papers* [1974,3] and [1984,2].

1912

1 A valószinűségszámítás néhány kérdéséről és bizonyos velök összefüggő határozott integrálokról (Über einige Fragen der Wahrscheinlichkeitsrechnung und gewisse damit zusammenhängende bestimmte Integrale), Dissertation, Budapest. [*Math. Phys. Lapok* 22 (1913), 53–73, 163–219.] (Compare [1913, 2].) (1)

2 A molecularis refractio (Über Molekularrefraktion), *Math. Phys. Lapok* 21, 155–60. (Compare [1913, 1]). (2)

3 Nyugalomban lévő oldat concentratio-megoszlása a gravitatiós térben (Konzentrationsverteilung einer ruhenden Lösung unter Einfluss der Gravitation), *Math. Phys. Lapok* 21, 170–72. (3)

4 Über ein Problem von Laguerre (with M. Fekete), *Rend. Circ. Mat. Palermo* 34, 89–120. (5)

5 Laguerre problémájáról (Über ein Problem von Laguerre), *Math. és Term. Ért.* 30, 783–96. (Compare [1912,4]). (6)

6 Sur un théorème de Stieltjes, *C. R. Acad. Sci. Paris* 155, 767–69. (7)

1913

1 Über Molekularrefraktion, *Phys. Z.* 13, 352–54. (4)

2 Berechnung eines bestimmten Integrals, *Math. Ann.* 74, 204–12. (8)

3 Sur un théorème de Laguerre, *C. R. Acad. Sci. Paris* 156, 996–99. (9)

4 Aequidistans ordinátákkal adott polynom valós gyökeiről (Über die reellen Nullstellen eines durch äquidistante Ordinaten bestimmten Polynoms), *Math. és Term. Ért.* 31, 438–47. (Compare [1913,3].) (10)

5 Über eine Peanosche Kurve, *Bull. Acad. Sci. Cracovie* A, 305–13. (12)

6 Sur la méthode de Graeffe, *C. R. Acad. Sci. Paris* 156, 1145–47. (13)

7 Sur un algorithme toujours convergent pour obtenir les polynomes de meilleure approximation de Tchebychef pour une fonction continue quelconque, *C. R. Acad. Sci. Paris* 157, 840–43. (14)

8 Über Annäherung durch Polynome mit lauter reellen Wurzeln, *Rend. Circ. Mat. Palermo* 36, 279–95. (16)

9 Über Annäherung durch Polynome deren sämtliche Wurzeln in einen Winkelraum fallen, *Nachr. Ges. Wiss. Göttingen*, 326–30. (17)

1914

1 Über einige Verallgemeinerungen der Descartesschen Zeichenregel, *Arch. Math. Phys.* 23 (3), 22–32. (11)

2 Über eine von Herrn C. Runge behandelte Integralgleichung, *Math. Ann.* 75, 376–79. (15)

3 Über einen Zusammenhang zwischen der Konvergenz von Polynomfolgen und der Verteilung ihrer Wurzeln (with E. Lindwart), *Rend. Circ. Mat. Palermo* 37 (1914), 297–304. (18)

4 Über das Graeffesche Verfahren, *Z. Math. Phys.* 63, 275–90. (19)

5 Über zwei Arten von Faktorenfolgen in der Theorie der algebraischen Gleichungen (with I. Schur), *J. Reine Angew. Math.* 144, 89–113. (20)

6 Sur une question concernant les fonctions entières, *C. R. Acad. Sci. Paris* 158, 330–33. (21)

7 Positiv quadratikus alakokról, a melyeknek mátrixa Hankel-féle (Über positive quadratische Formen mit Hankelscher Matrix), *Math. és Term. Ért.* 32, 656–79. (23)

1915

1 Algebraische Untersuchungen über ganze Funktionen vom Geschlechte Null und Eins, *J. Reine Angew. Math.* 145, 224–49. (22)

2 Über ganzwertige ganze Funktionen, *Rend. Circ. Mat. Palermo* 40, 1–16. (24)

3 Az algebrai egyenletek elméletéhez (Zur Theorie der algebraischen Gleichungen) *Math. és Term. Ért.* 33, 139–41. (Compare [1916,7].) (25)

4 Bemerkung zur Theorie der ganzen Funktionen, *Jahresber. Deutsch. Math.-Verein.* 24, 392–400. (28)

5 Une série de puissances est-elle en général non-continuable?, *Enseignement Math.* 17, 343–44.

1916

1 Zwei Beweise eines von Herrn Fatou vermuteten Satzes (with A. Hurwitz), *Acta Math.* 40, 179–83. (26)

2 Über den Zusammenhang zwischen dem Maximalbetrage einer analytischen Funktion und dem grössten Gliede der zugehörigen Taylorschen Reihe, *Acta Math.* 40, 311–19. (27)

3 Über Potenzreihen mit ganzzahligen Koeffizienten, *Math. Ann.* 77, 497–513. (29)

4 Eisenstein tételéről (Über den Eisensteinschen Satz), *Math. és Term. Ért.* 34, 754–58. (30)

5 Über das Anwachsen von ganzen Funktionen, die einer Differentialgleichung genügen, *Vierteljschr. Naturforsch. Ges. Zürich* 61, 531–45. (31)

6 Über algebraische Gleichungen mit nur reellen Wurzeln, *Vierteljschr. Naturforsch. Ges. Zürich* 61, 546–48. (32)

1917

1 Über eine neue Weise bestimmte Integrale in der analytischen Zahlentheorie zu gebrauchen, *Nachr. Ges. Wiss. Göttingen,* 149–59. (34)

2 Über geometrische Wahrscheinlichkeiten, *S.-B. Akad. Wiss. Wien* 126, 319–28. (36)

3 Généralisation d'un théorème de M. Størmer, *Arch. Math. Naturvid.* 35 (5), 8 pp. (37)

4 Über geometrische Wahrscheinlichkeiten an konvexen Körpern, *Ber. Sächs. Ges. Wiss. Leipzig* 69, 457–58. (38)

5 Un pendant du théorème d'approximation de Liouville dans la théorie des équations différentielles, *Enseignement Math.* 19, 96-97.

6 Anwendung des Riemannschen Integralbegriffes auf einige zahlentheoretische Aufgaben, *Arch. Math. Phys.* 26 (3), 196–201. (33)

7 Über die Potenzreihen, deren Konvergenzkreis natürliche Grenze ist, *Acta Math.* 41, 99–118. (35)

8 Sur les propriétés arithmétiques des séries entières, qui représentent des fonctions rationnelles, *Enseignement Math.* 19, 323. (See [1918, 8].)

9 Sur les polynomes à valeurs entiers dans un corps algébrique (with A. Ostrowski), *Enseignement Math.* 19, 323–24. (See [1918, 9].)

1918

1 Über Potenzreihen mit endlich vielen verschiedenen Koeffizienten, *Math. Ann.* 78, 286–93. (39)

2   Zur arithmetischen Untersuchung der Polynome, *Math. Z.* 1, 143–48. (40)

3   Über die Verteilung der quadratischen Reste und Nichtreste, *Nachr. Ges. Wiss. Göttingen,* 21–29. (41)

4   Über die Nullstellen gewisser ganzer Funktionen, *Math. Z.* 2, 352–83. (42)

5   Zahlentheoretisches und Wahrscheinlichkeitstheoretisches über die Sichtweite im Walde, *Arch. Math. Phys.* 27 (3), 135–42. (43)

6   Über die "doppelt-periodischen" Lösungen des *n*-Damen-Problems, Aus W. Ahrens, *Math. Unterhalt. Spiele,* 2-te Aufl., Teubner, Berlin, Vol. 2, pp. 364–74. (44)

7   Über die Verteilungssysteme der Proportionalwahl, *Z. Schweiz. Stat. Volkswirtsch.* 54, 363–87. (52)

8   Über arithmetische Eigenschaften der Reihenentwicklungen rationaler Funktionen, *Verh. Schweiz. Naturforsch. Ges.* 99, 123. (See [1920, 1].)

9   Über ganzwertige Polynome in algebraische Zahlkörpern (with A. Ostrowski), *Verh. Schweiz. Naturforsch. Ges.* 99, 136–37. (See [1919, 1].)

1919

1   Über ganzwertige Polynome in algebraischen Zahlkörpern, *J. Reine Angew. Math.* 149, 97–116. (45)

2   Verschiedene Bermerkungen zur Zahlentheorie, *Jahresber. Deutsch. Math.-Verein.* 28, 31–40. (48)

3   Zur Statistik der sphärischen Verteilung der Fixsterne, *Astr. Nachr.* 208, 175–80. (49)

4   Über das Gauss'sche Fehlergesetz, *Astr. Nachr.* 208, 185–92; 209, 111–12. (50)

5   Geometrische Darstellung einer Gedankenkette, *Schweiz. Pädagog. Z.,* 11 pp. (51)

6   Anschauliche und elementare Darstellung der Lexisschen Dispersionstheorie, *Z. Schweiz. Stat. Volkswirtsch.* 55, 121–40. (53)

7   Proportionalwahl und Wahrscheinlichkeitsrechnung, *Z. Ges. Staatswiss.* 74, 297–322. (55)

8   Sur la représentation proportionnelle en matière électorale, *Enseignement Math.* 20, 355–79. (56)

9   Wahrscheinlichkeitstheoretisches über die "Irrfahrt", *Mitt. Phys. Ges. Zürich* 19, 75–86. (58)

10 Das wahrscheinlichkeitstheoretische Schema der Irrfahrt, *Verh. Schweiz. Naturforsch. Ges.* 100, 76. (See [1921, 7].)

11 Über Sitzverteilung bei Proportionalwahlverfahren, *Schweiz. Zentralblatt f. Staats u. Gemeinde-Verwaltung* 20 (1), 1–5. (52a)

12 Über die Systeme der Sitzverteilung bei Proportionalwahl, *Wissen und Leben* 12, 259–68, 307–12.

13 Quelques problèmes de probabilité se rapportant à la "promenade au hasard", *Enseignement Math.* 20, 444–45. (See [1919, 10].)

1920

1 Arithmetische Eigenschaften der Reihenentwicklungen rationaler Funktionen, *J. Reine Angew. Math.* 151, 1–31. (46)

2 Zur Untersuchung der Grössenordnung ganzer Funktionen, die einer Differentialgleichung genügen, *Acta Math.* 42, 309–16. (47)

3 Über den zentralen Grenzwertsatz der Wahrscheinlichkeitsrechnung und das Momentenproblem, *Math. Z.* 8, 171–81. (54)

4 Über ganze ganzwertige Funktionen, *Nachr. Ges. Wiss. Göttingen*, 1–10. (57)

5 Geometrisches über die Verteilung der Nullstellen gewisser ganzer transzendenter Funktionen, *S.-B. Bayer. Akad. Wiss.*, 285–90. (59)

6 Sur les fonctions entières, *Enseignement Math.* 21, 217–18.

1921

1 Anschaulich-experimentelle Herleitung der Gaußschen Fehlerkurve, *Z. Math. Naturwiss. Unterricht* 52, 57–65. (60)

2 Neuer Beweis für die Produktdarstellung der ganzen transzendenten Funktionen endlicher Ordnung, *S.-B. Bayer. Akad. Wiss. Math. Nat. Abteilung.*, 29–40. (61)

3 En funktionsteoretisk bemaerkning, *Mat. Tidsskr.*, B, 14–16. (62)

4 Bestimmung einer ganzen Funktion endlichen Geschlechts durch viererlei Stellen, *Mat. Tidsskr.*, B, 16–21. (63)

5 Ein Mittelwertsatz für Funktionen mehrerer Veränderlichen, *Tôhoku Math. J.* 19, 1–3. (64)

6 Über die kleinsten ganzen Funktionen, deren sämtliche Derivierten im Punkte $z = 0$ ganzzahlig sind, *Tôhoku Math. J.* 19, 65–68. (65)

7 Über eine Aufgabe der Wahrscheinlichkeitsrechnung betreffend die Irrfahrt im Straßennetz, *Math. Ann.* 84, 149–60. (66)

8 Bemerkung über die Mittag-Lefflerschen Funktionen $E_\alpha(z)$, *Tôhoku Math. J.* 19, 241–48. (71)

9 Eine Ergänzung zu dem Bernoullischen Satz der Wahrscheinlichkeitsrechnung, *Nach. Ges. Wiss. Göttingen*, 223–28; (1923), 96. (70)

1922

1 Über die Nullstellen sukzessiver Derivierten, *Math. Z.* 12, 36–60. (67)

2 Sur les séries entières dont la somme est une fonction algébrique, *Enseignement Math.* 22, 38–47. (68)

3 Arithmetische Eigenschaften und analytischer Charakter, *Jahresber. Deutsch. Math.-Verein.* 30, 88–89; 31, 107–15. (72)

4 Über eine arithmetische Eigenschaft gewisser Reihenentwicklungen, *Tôhoku Math. J.* 22, 79–81. (73)

5 Prolongement analytique, *Enseignement Math.* 22, 298–99.

6 Sur les zéroes des dérivées successives, *Enseignement Math.* 22, 68. (See [1922, 1].)

1923

1 Sur les séries entières à coefficients entiers, *Proc. London Math. Soc.* 21 (2), 22–38. (69)

2 Herleitung des Gaußschen Fehlergesetzes aus einer Funktionalgleichung, *Math. Z.* 18, 96–108. (74)

3 Bemerkungen über unendliche Folgen und ganze Funktionen, *Math. Ann.* 88, 169–83. (75)

4 Analytische Fortsetzung und konvexe Kurven, *Math. Ann.* 89, 179–91. (76)

5 Über die Existenz unendlich vieler singulärer Punkte auf der Konvergenz-geraden gewisser Dirichletscher Reihen, *S.-B. Preuß. Akad., Phys.-Math. Kl.* 5, 45–50. (77)

6 Rationale Abzählung der Gitterpunkte (with R. Fueter), *Vierteljschr. Naturforsch. Ges. Zürich* 68, 380–86. (78)

7 Über die Statistik verketteter Vorgänge (with F. Eggenberger), *Z. Angew. Math. Mech.* 3, 279–89. (79)

8 On the zeros of an integral function represented by Fourier's integral, *Messenger of Math.* 52, 185–88. (81)

1924

1 Über die Analogie der Krystallsymmetrie in der Ebene, *Z. Krystall.* 60, 278–82. (82)

2 Sur certaines transformations fonctionnelles linéaires des fonctions analytiques, *Bull. Soc. Math. France* 52, 519–32. (83)

3 On the mean-value theorem corresponding to a given linear homogeneous differential equation, *Bull. Amer. Math. Soc.* 30, 10. (Compare [1924,4].)

4 On the mean-value theorem corresponding to a given linear homogeneous differential equation, *Trans. Amer. Math. Soc.* 24 (1922), 312–24. (80)

1925

1 Sur l'existence d'une limite considérée par M. Hadamard, *Enseignement Math.*, 24, 76–78. (84)

2 Über eine geometrische Darstellung der Fareyschen Reihe, *Acta Lit. Sci. Szeged* 2, 129–33. (85)

3* *Aufgaben und Lehrsätze aus der Analysis.* I. *Reihen, Integralrechnung, Funktionentheorie*; II. *Funktionentheorie, Nullstellen, Polynome, Determinanten, Zahlentheorie* (with G. Szegő), Springer. (New editions in 1945, 1954, 1964, 1970–71. Translations: Bulgarian, 1973; English, 1972, 1976, 1998; Hungarian, 1980, 1981; Russian, 1978.)

4* Wahrscheinlichkeitsrechnung, Fehlerausgleichung, Statistik, *Abderhalden's Handbuch der biologischen Arbeitsmethoden*, Vol. 2, pp. 669-758.

1926

1 On certain sequences of polynomials, *Proc. London Math. Soc.* 24 (2), xliv–xlv. (86)

2 Proof of an inequality, *Proc. London Math. Soc.* 24 (2), lvii. (87)

3 On an integral function of an integral function, *J. London Math. Soc.* 1, 12–15. (88)

4 On the minimum modulus of integral functions of order less than unity, *J. London Math. Soc.* 1, 78–86. (89)

5 On the zeros of certain trigonometric integrals, *J. London Math. Soc.* 1, 98–99. (90)

6 The maximum of a certain bilinear form (with G. H. Hardy and J. E. Littlewood), *Proc. London Math. Soc.* 25 (2), 265–82. (91)

7 Application of a theorem connected with the problem of moments, *Messenger of Math.* 55, 189–92. (92)

8 Bemerkung über die Integraldarstellung der Riemannschen ξ-Funktion, *Acta Math.* 48, 305–17. (93)

9 Sopra una equazione trascendente trattata da Eulero, *Boll. Un. Mat. Ital.* 5, 64–68. (94)

10 Sur les opérations fonctionnelles linéaires échangeables avec la dérivation et sur les zéros des polynomes, *C. R. Acad. Sci. Paris* 183, 413–14. (95)

11 Sur les opérations fonctionnelles linéaires échangeables avec la dérivation et sur les zéros des sommes d'exponentielles, *C. R. Acad. Sci. Paris* 183, 467–68. (96)

1927

1 Sur une condition à laquelle satisfont les coefficients des séries entières prolongeables, *Enseignement Math.* 26, 313–14.

2 Sur les singularités des séries lacunaires, *C. R. Acad. Sci. Paris* 184, 502–4. (97)

3 Sur un théorème de M. Hadamard relatif à la multiplication des singularités, *C. R. Acad. Sci. Paris* 184, 579–81. (98)

4 Theorems concerning mean values of analytic functions (with G. H. Hardy and A. E. Ingham), *Proc. Roy. Soc.* A 113, 542–69. (99)

5 Sur les fonctions entières à série lacunaire, *C. R. Acad. Sci. Paris* 184, 1526–28. (100)

6 Über trigonometrische Integrale mit nur reellen Nullstellen, *J. Reine Angew. Math.* 158, 6–18. (101)

7 Über die algebraisch-funktionentheoretischen Untersuchungen von J. L. W. V. Jensen, *Kgl. Danske Vid. Sel. Math.-Fys. Medd.* 7 (17), 33 pp. (102)

8 Note on series of positive terms, *J. London Math. Soc.* 2, 166–69. (103)

9 Elementarer Beweis einer Thetaformel, *S.-B. Preuß. Akad., Phys.-Math. Kl.* 9, 158–61. (104)

10 Sur les coefficients de la série de Taylor, *C. R. Acad. Sci. Paris* 185, 1107–8. (105)

11 Eine Verallgemeinerung des Fabryschen Lückensatzes, *Nachr. Ges. Wiss. Göttingen,* 187–95. (106)

12 Notwendige Determinantenkriterien für die Fortsetzbarkeit einer Potenzreihe, *Verh. Schweiz. Naturforsch. Ges.* 84–85. (See [1928, 43].)

1928

1 Über positive Darstellung von Polynomen, *Vierteljschr. Naturforsch. Ges. Zürich* 73, 141–45. (107)

2 Notes on moduli and mean values (with G. H. Hardy and A. E. Ingham), *Proc. London Math. Soc.* 27 (2), 401–9. (108)

3 Über die Funktionalgleichung der Exponentialfunktion im Matrizenkalkül, *S.-B. Preuß. Akad. Phys.-Math. Kl.* 10, 96–99. (109)

4 Über gewisse notwendige Determinantenkriterien für die Fortsetzbarkeit einer Potenzreihe, *Math. Ann.* 99, 687–706. (110)

5 Beitrag zur Verallgemeinerung des Verzerrungssatzes auf mehrfach zusammenhängende Gebiete, *S.-B. Preuß. Akad., Phys.-Math. Kl.* (a), 228–32; (b) second part, 280–82 (1928); (c) third part (1929), 55–62. (111)

6 Sur l'interprétation de certaines courbes de fréquence (with F. Eggenberger), *C. R. Acad. Sci. Paris* 187, 870–72. (112)

7 Sur la recherche des points singuliers de la série de Taylor, *Atti del Congresso Internazionale, Bologna,* Vol. III, 243–47. (122)

8 Über eine Eigenschaft des Gaußschen Fehlergesetzes, *Atti del Congresso Internazionale, Bologna,* Vol. VI, 63–64. (131)

1929

1 Untersuchungen über Lücken und Singularitäten von Potenzreihen, *Math. Z.* 29, 549–640. (113)

2 Über einem Satz von Laguerre, *Jahresber. Deutsch. Math.-Verein.* 38, 161–68. (114)

3 Untersuchungen über Lücken und Singularitäten von Potenzreihen, *Boll. Un. Mat. Ital.* 8, 211–14. (Résumé of [1929, 1].)

1930

1 Some simple inequalities satisfied by convex functions (with G. H. Hardy and J. E. Littlewood), *Messenger of Math.* 58, 145–52. (115)

2 Eine Wahrscheinlichkeitsaufgabe in der Kundenwerbung, *Z. Angew. Math. Mech.* 10, 96–97. (116)

3 Über das Vorzeichen des Restgliedes im Primzahlsatz, *Nachr. Ges. Wiss. Göttingen,* [Fachgr. 1, no. 2], 19–27. (117)

4 Some problems connected with Fourier's work on transcendental equations, *Quart. J. Math., Oxford Ser.* 1, 21–34. (118)

5 Eine Wahrscheinlichkeitsaufgabe in der Pflanzensoziologie, *Vierteljschr. Naturforsch. Ges. Zürich* 75, 211–19. (119)

6 Liegt die Stelle der grössten Beanspruchung an der Oberfläche, *Z. Angew. Math. Mech.* 10, 353–60. (120)

1931

1 Sur quelques points de la théorie des probabilités, *Ann. Inst. H. Poincaré* 1, 117–61. (121)

2 Über Potenzreihen mit ganzen algebraischen Koeffizienten, *Abh. Math. Sem. Univ. Hamburg* 8, 401–2. (123)

3 Sur quelques propriétés qualitatives de la propagation de la chaleur (with G. Szegő), *C. R. Acad. Sci. Paris* 192, 1340–41. (124)

4 Über den transfiniten Durchmesser (Kapazitätskonstante) von ebenen und räumlichen Punktmengen (with G. Szegő), *J. Reine Angew. Math.* 165, 4–49. (125)

5 Sur les valeurs moyennes des fonctions réelles définies pour toutes les valeurs de la variable (with M. Plancherel), *Comment. Math. Helv.* 3, 114–21. (126)

6 Bemerkung zur Interpolation und zur Näherungstheorie der Balkenbiegung, *Z. Angew. Math. Mech.* 11, 445–49. (129)

7 Comment chercher la solution d'un problème de mathématiques? *Enseignement Math.* 30, 275–76. (Résumé of [1932,3].) (130A)

1932

1 On polar singularities of power series and of Dirichlet series, *Proc. London Math. Soc.* 33 (2), 85–101. (127)

2   On the roots of certain algebraic equations (with A. Bloch), *Proc. London Math. Soc.* 33 (2), 102–14. (128)

3   Wie sucht man die Lösung mathematischer Aufgaben? *Z. Math. Naturwiss. Unterricht* 63, 159–69. (130)

4   Über einen Satz von Myrberg, *Jahresber. Deutsch. Math.-Verein.* 42, 159. (132)

5* *Mathematische Werke von Adolf Hurwitz.* I. *Funktionentheorie.* II. *Zahlentheorie, Algebra, und Geometrie* (ed. G. Pólya), Birkhäuser, 1932, 1933.

1933

1   Abschätzung des Betrages einer Determinante (with A. Bloch), *Vierteljschr. Naturforsch. Ges. Zürich* 78, 27–33. (133)

2   Über die Konvergenz von Quadraturverfahren, *Math. Z.* 37, 264–86. (134)

3   Qualitatives über Wärmeausgleich, *Z. Angew. Math. Mech.* 13, 125–28. (135)

4   Über analytische Deformationen eines Rechtecks, *Ann. Math.* 34 (2), 617–20. (136)

5   Untersuchungen über Lücken and Singularitäten von Potenzreihen, Zweite Mitteilung. (Continuation of [1929, 1]), *Ann. Math.* 34 (2), 731–77. (137)

6   Bemerkung zu der Lösung der Aufgabe l05, *Jahresber. Deutsch. Math.-Verein.* 43, 67–69. (138)

7   Review of *Distribution of Primes* by A. E. Ingham, *Math. Gaz.* 17, 329–30.

1934

1   Quelques théorèmes analogues au théorème de Rolle, liés à certaines équations linéaires aux dérivées partielles, *C. R. Acad. Sci. Paris* 199, 655–57. (139)

2   Sur l'application des opérations différentielles linéaires aux séries, *C. R. Acad. Sci. Paris* 199, 766–67. (140)

3   Über die Potenzreihenentwicklung gewisser mehrdeutiger Funktionen, *Comment. Math. Helv.* 7, 201–21. (141)

4* *Inequalities* (with G. H. Hardy and J. E. Littlewood), Cambridge University Press. (New editions, 1952, 1988. Translations: Chinese, 1965; Russian, 1948.)

1935

1   On the power series of an integral function having an exceptional value (with A. Pfluger), *Proc. Cambridge Philos. Soc.* 31, 153–55. (142)

2   Zwei Aufgaben aus der Wahrscheinlichkeitsrechnung, *Vierteljschr. Naturforsch. Ges. Zürich* 80, 123–30. (143)

3   Sur les séries entières satisfaisant à une équation différentielle algébrique, *C. R. Acad. Sci. Paris* 201, 444–45. (144)

4 Un problème combinatoire général sur les groupes de permutations et le calcul du nombre des isomères des composés organiques, *C. R. Acad. Sci. Paris* 201, 1167–70. (145)

1936

1 Tabelle der Isomerenzahlen für die einfacheren Derivate einiger cyclischen Stammkörper, *Helv. Chim. Acta* 19, 22–24. (146)
2 Algebraische Berechnung der Anzahl der Isomeren einiger organischer Verbindungen, *Z. Kristall.* (A), 93, 415–43. (147)
3 Sur le nombre des isomères de certains composés chimiques, *C. R. Acad Sci. Paris* 202, 1554–56. (148)
4 Über das Anwachsen der Isomerenzahlen in den homologen Reihen der organischen Chemie, *Vierteljschr. Naturforsch. Ges. Zürich* 81, 243–58. (149)
5 (Announcement) *Comptes Rendus, Congrès International d. Math. Oslo,* Vol. 2, 19. (See [1937,3].) (152A)

1937

1 Zur Kinematik der Geschiebebewegung, *Mitt. Vers. Wass. E. T. H. Zürich* 1–21. (150)
2 Fonctions entières et intégrales de Fourier multiples (with M. Plancherel), *Comment. Math. Helv.* (a) Erste Mitteilung, 9 (1936/37), 224–48; (b) Zweite Mitteilung, 10 (1937/38), 110–63. (151)
3 Kombinatorische Anzahlbestimmungen für Gruppen, Graphen und chemische Verbindungen, *Acta Math.* 68, 145–254. (152)
4 Über die Realität der Nullstellen fast aller Ableitungen gewisser ganzer Funktionen, *Math. Ann.* 114, 622–34. (153)
5 Le problème de Tours, *Sphinx* 1, 108–10.

1938

1 Sur l'indétermination d'un problème voisin du problème des moments, *C. R. Acad. Sci. Paris* 207, 708–11. (154)
2 Sur la promenade au hasard dans un réseau de rues, *Actualités Sci. Ind.* 734, 25–44. (155)
3 Wie sucht man die Lösung mathematischer Aufgaben? *Acta Psych.* 4, 113–70. (156)
4 Eine einfache, mit funktionentheoretischen Aufgaben verknüpfte, hinreichende Bedingung für die Auflösbarkeit eines Systems unendlich vieler linearer Gleichungen, *Comment. Math. Helv.* 11 (1938/39), 234–52. (158)

1939

1 Sur les séries entières lacunaires non prolongeables, *C. R. Acad. Sci. Paris* 208, 709–11. (157)

1940

1 Sur les types des propositions composées, *J. Symbolic Logic* 5, 98–103. (159)

1941

1 On functions whose derivatives do not vanish in a given interval, *Proc. Nat. Acad. Sci.* 27, 216–17. (160)

2 Generalizations of completely convex functions (with R. P. Boas, Jr.), *Proc. Nat. Acad. Sci.* 27, 323–25. (161)

3 Sur l'existence de fonctions entières satisfaisant à certaines conditions linéaires, *Trans. Amer. Math. Soc.* 50, 129–39. (162)

4 Heuristic reasoning and the theory of probability, *Amer. Math. Monthly* 48, 450–65. (163)

1942

1 On converse gap theorems, *Trans. Amer. Math. Soc.* 52, 65–71. (164)

2 Influence of the signs of the derivatives of a function on its analytic character (with R. P. Boas, Jr.), *Duke Math. J.* 9, 406–24. (165)

3 On the oscillation of the derivatives of a periodic function (with N. Wiener), *Trans. Amer. Math. Soc.* 52, 249–56. (166)

1943

1 On the zeros of the derivatives of a function and its analytic character, *Bull. Amer. Math. Soc.* 49, 178–91. (167)

2 Approximations to the area of the ellipsoid, *Publ. Inst. Mat. Rosario* 5, 13 pp. (168)

1945

1 Inequalities for the capacity of a condenser (with G. Szegő), *Amer. J. Math.* 67, 1–32. (169)

2* *How to Solve It: A New Aspect of Mathematical Method*, Princeton University Press. (New printings in 1946, 1948, 1954, 1971; Doubleday editions in 1957, 1958.

Translations: Arabic, 1960; Bulgarian, n.d.; Chinese, 1984; Dutch, 1974; French, 1957, 1962, 1965; German, 1949, 1957; Greek, 1991; Hebrew, 1961; Hungarian, 1957, 1969, 1971, 1977; Italian, 1967, 1976; Japanese, 1954; Malaysian, 1993; Polish, 1964, 1993; Portuguese, 1977; Romanian, 1965; Russian, 1959; Serbo-Croatian, 1956, 1966; Slovenian, 1976; Spanish, 1965; Swedish, 1970; Turkish, 1997.)

1946

1 Sur une généralisation d'un problème élémentaire classique, importante dans l'inspection des produits industriels, *C. R. Acad. Sci. Paris* 222, 1422–24. (170)

1947

1 Estimating electrostatic capacity, *Amer. Math. Monthly* 54, 201–06. (171)

2 A minimum problem about the motion of a solid through a fluid, *Proc. Nat. Acad. Sci.* 33, 218–21. (172)

3 Sur la fréquence fondamentale des membranes vibrantes et la résistance élastique des tiges à la torsion, *C. R. Acad. Sci. Paris* 225, 346–48. (173)

1948

1 On patterns of plausible inference, *Courant Anniversary Volume*, 277–88. (174)

2 Exact formulas in the sequential analysis of attributes, *University of California Publications in Mathematics* (New Series) 1, (5) 229–40. (175)

3 Generalization, specialization, analogy, *Amer. Math. Monthly* 55, 241–43. (176)

4 Torsional rigidity, principal frequency, electrostatic capacity and symmetrization, *Quart. Appl. Math.* 6, 267–77. (177)

1949

1 On the product of two power series (with H. Davenport), *Canad. J. Math.* 1, 1–5. (178)

2 Remarks on computing the probability integral in one and two dimensions, *Proceedings of the Berkeley Symposium on Mathematical Statistics and Probability*, University of California Press, 63–78. (179)

3 Remarks on characteristic functions, *Proceedings of the Berkeley Symposium on Mathematical Statistics and Probability*, University of California Press, 115–23. (180)

4 Sur les symétries des fonctions sphériques de Laplace (with Burnett Meyer), *C. R. Acad. Sci. Paris* 228, 28–30. (181)

5   Sur les fonctions sphériques de Laplace de symétrie cristallographique donnée (with Burnett Meyer), *C. R. Acad. Sci. Paris* 228, 1083–84. (182)

6   Preliminary remarks on a logic of plausible inference, *Dialectica* 3, 28–35. (183)

7   With, or without, motivation?, *Amer. Math. Monthly* 56, 684–91. (184)

8   Statement concerning the article by N. Wiener on G. H. Hardy, *Bull. Amer. Math. Soc.* 55, 1082.

9   On solving mathematical problems in high school, *Calif. Math. Council Bull.* 7, (2) 3, 17.

1950

1   Remark on Weyl's note "Inequalities between the two kinds of eigenvalues of a linear transformation", *Proc. Nat. Acad. Sci.* 36, 49–51. (185)

2   On the harmonic mean of two numbers, *Amer. Math. Monthly* 57, 26–28. (186)

3   Sur la symétrisation circulaire, *C. R. Acad. Sci. Paris* 230, 25–27. (187)

4   Remarks on power series, *Acta Sci. Math.* 12B, 199–203; 14 (1951–52), 144. (188)

5   On the torsional rigidity of multiply connected cross-sections (with Alexander Weinstein), *Ann. Math.* 52, 154–63. (189)

6   Let us teach guessing, *Études de Philosophie des Sciences, en hommage à Ferdinand Gonseth,* Neuchâtel: Griffon, pp. 147–54. (190)

7   On plausible reasoning, *Proceedings of the International Congress of Mathematicians,* Providence: Amer. Math. Soc., Vol. I, 739–47. (192)

1951

1   A note on the principal frequency of a triangular membrane, *Quart. Appl. Math.* 8, 386. (191)

2* *Isoperimetric Inequalities in Mathematical Physics* (with G. Szegő), Princeton University Press. (Translation: Russian, 1962.)

1952

1   Remarks on the foregoing paper (by E. T. Kornhauser and I. Stakgold), *J. Math. Phys.* 31, 55–57. (193)

2   Remarques sur un problème d'algèbre étudié par Laguerre, *J. Math. Pures Appl.* 31, (9) 37–47. (194)

3   Sur une interprétation de la méthode des différences finies qui peut fournir des bornes supérieures ou inférieures, *C. R. Acad. Sci. Paris* 235, 995–97. (195)

4 Sur le rôle des domaines symétriques dans le calcul de certaines grandeurs physiques, *C. R. Acad. Sci. Paris* 235, 1079–81. (196)

1953

1 *Convexity of functionals by transplantation* (with M. Schiffer), U.S. Office of Naval Research Technical Report 14. (Compare [1954, 1].)

2 Great and small examples of problem solving, *Proceedings of a Summer Conference in Collegiate Mathematics,* University of Colorado, 48 pp.

3 *Two notes on minimum principle approximations,* Technical Report 29, Stanford University.

4 The 1953 Stanford Competitive Examination: problems, solutions, and comments, *Calif. Math. Council Bull.* (1) 11, 15–17.

1954

1 Convexity of functionals by transplantation (with M. Schiffer), *J. Analyse Math.* 3, 245–345. (197)

2 An elementary analogue to the Gauss-Bonnet theorem, *Amer. Math. Monthly* 61, 601–3. (198)

3 Estimates for eigenvalues, *Studies in Mathematics and Mechanics presented to Richard von Mises,* Academic Press, pp. 200–7. (199)

4* *Mathematics and Plausible Reasoning.* I. *Induction and Analogy in Mathematics.* II. *Patterns of Plausible Inference,* Princeton University Press. (Second edition of II, 1968. Translations: Bulgarian, 1970; French, 1957, 1958; German, 1962, 1963; Hungarian, 1987; Japanese, 1959; Romanian, 1962; Russian, 1957, 1975; Spanish, 1966; Turkish, 1966.)

5 The 1954 Stanford Competitive Examination: problems and solutions, *Calif. Math. Council Bull.* (2) 12, 7–8.

1955

1 More isoperimetric inequalities proved and conjectured, *Comment. Math. Helv.* 29, 112–19. (200)

2 *On the characteristic frequencies of a symmetric membrane,* Applied Mathematics and Statistics Technical Report 40, Stanford University. (Compare [1955, 4].)

3 Sur le quotient de deux fréquences propres consécutives (with L. E. Payne and H. F. Weinberger), *C. R. Acad. Sci. Paris* 241, 917–19. (201)

4 On the characteristic frequencies of a symmetric membrane, *Math. Z.* 63, 331–37. (202)

5 *Estimates for eigenvalues,* Applied Mathematics and Statistics Technical Report 34, Stanford University. (Compare [1954, 3].)

6 *More isoperimetric inequalities proved and conjectured,* Applied Mathematics and
   Statistics Technical Report 38, Stanford University. (Compare [1955, 1].)
7 The 1955 Stanford Competitive Examination in Mathematics: problems, solutions, and comments, *Calif. Math. Council Bull.* (2) 13, 15–17.

1956

1 On the ratio of consecutive eigenvalues (with L. E. Payne and H. F.
   Weinberger), *J. Math. Phys.* 35, 289–98. (203)
2 Sur les fréquences propres des membranes vibrantes, *C. R. Acad. Sci. Paris*
   242, 708–9. (204)
3 On picture-writing, *Amer. Math. Monthly* 63, 689–97. (205)
4 Die Mathematik als Schule des plausiblen Schliessens, *Gymn. Helv.* 10, 4–8;
   Archimedes 8, 111–14; Mathematics as a subject for learning plausible reasoning (translation by C. M. Larsen), *Math. Teacher* 52 (1959), 7–9. (206A)
5 Sur quelques membranes vibrantes de forme particulière, *C. R. Acad. Sci.
   Paris* 243, 467–69. (207)
6 *On the eigenvalues of certain membranes (two notes),* Applied Mathematics and
   Statistics Laboratory Technical Report 58, Stanford University.
7 *On the ratio of consecutive eigenvalues* (with L. E. Payne and H. F. Weinberger),
   Applied Mathematics and Statistics Technical Report 41, Stanford University. (Compare [1956, 1].)
8 The 1956 Stanford Competitive Examination in Mathematics: solutions and
   comments, *Calif. Math. Council Bull.* (2) 14, 19–22.

1957

1 *Remarks on de la Vallée-Poussin means and conformal maps of the circle* (with I. J.
   Schoenberg), Stanford University Applied Mathematics and Statistics Laboratory Technical Report 70. (Compare [1958, 3].)
2 The 1957 Stanford University Competitive Examination in Mathematics, *Calif.
   Math. Council Bull.* (2) 15, 18–21.

1958

1 L'Heuristique est-elle un sujet d'étude raisonnable?, *La méthode dans les sciences modernes,* Travail et Méthode, numéro hors série, 279–85. (206)
2 On the curriculum for prospective high school teachers, *Amer. Math. Monthly*
   65, 101–4. (208)
3 Remarks on de la Vallée-Poussin means and convex conformal maps of the
   circle (with I. J. Schoenberg), *Pacific J. Math.* 8, 295–334. (209)

4 The 1958 Stanford-Sylvania University Competitive Examination in Mathematics, *Calif. Math. Council Bull.* (1) 16, 18–20.

1959

1 Ten Commandments for Teachers, *Journal of Education of the Faculty and College of Education of the University of British Columbia*; Vancouver and Victoria (3), 61–69. (210)

2 Sur la représentation conforme de l'extérieur d'une courbe fermée convexe (with M. Schiffer), *C. R. Acad. Sci. Paris* 248, 2837–39. (211)

3 Heuristic reasoning in the theory of numbers, *Amer. Math. Monthly* 66, 375–84. (212)

4 On the location of the centroid of certain solids (with C. J. Gerriets), *Amer. Math. Monthly* 66, 875–79. (213)

1960

1 Teaching of mathematics in Switzerland, *Amer. Math. Monthly* 67, 907–14; *Math. Teacher* 53, 552–58. (216)

2 Two more inequalities between physical and geometrical quantities, *J. Indian Math. Soc.* 24, 413–19. (219)

3 On the role of the circle in certain variational problems. In memoriam Lipót Fejér, *Ann. Univ. Sci. Budapest Eötvös Sect. Math.* 3–4 (1960/61), 233–39. (220)

4 Induktion und mathematische Induktion, *Matematyka* [Wrocław], 13, 283–88 (Polish).

5 Three puzzles and a pattern, *Math. Log* 3, (1) 1; 3, (3) 1, 5.

6 I. *On the eigenvalues of vibrating membranes;* II. *Two more inequalities between physical and geometrical quantities.* Applied Mathematics and Statistics Technical Report 88, Stanford University.

7 The 1960 Stanford-Sylvania Competitive Examination in Mathematics, *Calif. Math. Council Bull.* (2) 18, 16–17.

1961

1 Circle, sphere, symmetrization and some classical physical problems, *Modern mathematics for the engineer,* 2nd series, McGraw-Hill, pp. 420–41. (214)

2 On the eigenvalues of vibrating membranes, In memoriam Hermann Weyl, *Proc. London Math. Soc.* (3) 11, 419–33. (215)

3 The minimum fraction of the popular vote that can elect the President of the United States, *Math. Teacher* 54, 130–33. (217)

4 Leopold Fejér, *J. London Math. Soc.* 36, 501–6. (218)

5 The 1961 Stanford-Sylvania Competitive Examination in Mathematics, *Calif. Math. Council Bull.* (2) 19, 10–11.

1962

1* *Mathematical Discovery; on Understanding, Learning and Teaching Problem Solving,* two volumes, John Wiley and Sons, 1962, 1965. (Combined paperback edition, 1981. Translations: Bulgarian, 1968; French, 1967; German 1966, 1967, 1979, 1983; Hungarian, 1967, 1979; Italian, 1970-71, 1979, 1982; Japanese, 1964; Polish, 1975; Romanian, 1971; Russian, 1970, 1976.)

2 The teaching of mathematics and the biogenic law, *The scientist speculates* (ed. I. J. Good), Heinemann, London, pp. 352–56. (221)

1963

1 Intuitive outline of the solution of a basic combinatorial problem, *Switching theory in space technology* (ed. H. Aiken and W. F. Main), Stanford University Press, pp. 3–7. (222)

2 On learning, teaching and learning teaching, *Amer. Math. Monthly* 70, 605–19; *Readings in Secondary School Mathematics* (ed. D. Aichele, R. Reys), Prindle, Weber & Schmidt, 1971; *Neue Sammlung Göttinger Blätter für Kultur und Erziehung* 4 (3) (1964), 194–210; *Teaching and Learning: a Problem-Solving Focus,* National Council of Teachers of Mathematics, 1986. (223)

3* *Mathematical Methods in Science* (ed. Leon Bowden), School Mathematics Study Group. (New Edition, [1977, 1]. Translations: Hungarian, 1977, 1984; Italian, 1979.)

1964

1 Introduction, *Applied Combinatorial Mathematics* (ed. E. F. Beckenbach), John Wiley and Sons, pp. 1–2. (224)

2 Die Heuristik; Versuch einer vernünftigen Zielsetzung; Vermuten und wissenschaftliche Methode, *Die Mathematikunterricht* 10, (1) 5–15, 80–96. (See [1958, 1].)

1966

1 A series for Euler's constant, *Research Papers in Statistics, Festschrift for J. Neyman,* ed. F. N. David, John Wiley and Sons, pp. 259–61. (225)

2 On teaching problem solving, *The Role of Axiomatics and Problem Solving in Mathematics,* Ginn, pp. 123–29. (226)

3 A note of welcome, *J. Combinatorial Theory* 1, 1–2. (227)

1967

1   L'enseignement par les problèmes, *Enseignement Math.*, Sér. 2, 13, 233–41. (226A)

2   Inequalities and the principle of nonsufficient reason, *Inequalities* (ed. Oved Shisha), Academic Press, pp. 1–15. (228)

3   Introduction, Training in Applied Mathematics Research, *SIAM Rev.*, 9, 347; *Education in Applied Mathematics, Proceedings of a Conference, University of Denver,* Aspen, Colorado, May 24–27, 1966.

1968

1   Graeffe's method for eigenvalues, *Numer. Math.* 11, 315–19. (229)

2   Über das Vorzeichen des Restgliedes in Primzahlsatz, *Abhandlungen aus Zahlentheorie und Analysis; zur Erinnerung an Edmund Landau* (P. Turán), Plenum, Berlin, pp. 233–44.

3   Preface, *The Prime Imperatives/Priorities in Education* by Alexander Israel Wittenberg, Clarke, Irwin and Company Limited, Toronto and Vancouver, pp. v–vi.

1969

1   On the number of certain lattice polygons, *J. Combinatorial Theory* 6, 102–5. (230)

2   Fundamental ideas and objectives of mathematical education, *Mathematics in Commonwealth Schools*, pp. 27–34. (231)

3   Entiers algébriques, polygones et polyèdres réguliers, *Enseignement Math.* Sér 2, 15, 237–43. (232)

4   Some mathematicians I have known, *Amer. Math. Monthly* 76, 746–53; *Pokroky Mat. Fysiky a Astron.* 17 (1972) 237–44 (Czech); *Fiz.-Mat. Spis. Bulgar. Akad. Nauk.* (46) 13 (1970), 123–30 (Bulgarian). (233)

5   On the isoperimetric theorem: history and strategy, *Mathematical Spectrum* 2, 5–7. (234)

1970

1   Gaussian binomial coefficients and enumeration of inversions, *Proceedings of the Second Chapel Hill Conference on Combinatorial Mathematics and Its Applications,* University of North Carolina, Chapel Hill, N.C., pp. 381–84. (235)

2   Two incidents, *Scientists at work: Festschrift in Honour of Herman Wold* (ed. T. Dalenius, G. Karlsson and S. Malmquist), Almqvist & Wiksells Boktryckeri AB, Uppsala, Sweden, pp. 165–68. (236)

1971

1 Gaussian binomial coefficients (with G. L. Alexanderson), *Elem. Math.* 26, 102–9. (237)

2 Methodology or heuristics, strategy or tactics?, *Arch. Philos.* 34, 623–29. (238)

1972

1 Eine Erinnerung an Hermann Weyl, *Math. Z.* 126, 296–98. (239)

2 Formation, not only information, *Graduate Training of Mathematics Teachers,* Canadian Mathematical Congress, Montreal, pp. 53–62. (243)

1973

1 A letter by Professor Pólya, *Amer. Math. Monthly* 80, 73–74. (240)

2 As I read them, *Developments in Mathematical Education, Proceedings of the 2nd International Congress in Mathematical Education* (ed. A. G. Howson), Cambridge University Press, pp. 77–78. (244)

3 A story with a moral, *Math. Gazette* 57, 86–87. (242)

4 The Stanford University Competitive Examination in Mathematics (with J. Kilpatrick), *Amer. Math. Monthly* 80, 627–40. (241)

1974

1* *Complex Variables* (with G. Latta), John Wiley and Sons. (Translation: Spanish, 1976.)

2* *The Stanford Mathematics Problem Book/with Hints and Solutions* (with J. Kilpatrick), Teacher's College Press, New York. (Problems and solutions had earlier been published in the *Amer. Math. Monthly* 1946–1953, and the *Calif. Math. Council Bull.* 1953–1961.)

3* *George Pólya: Collected Papers.* Vol. I. *Singularities of Analytic Functions* (ed. R. P. Boas, Jr.). Vol. II. *Location of Zeros* (ed. R. P. Boas, Jr.). MIT Press. (See [1984, 2].)

1975

1 Partitions of a finite set into structured subsets, *Math. Proc. Camb. Phil. Soc.* 77, 453–58. (245)

1976

1 Probabilities in proofreading, *Amer. Math. Monthly* 83, 42. (246)

2 Guessing and proving, *California Math.* 1, 1–8; *Two-Year College Math. J.* 9 (1978), 21–27. (247)

3   On the zeros of successive derivatives, an example, *J. Analyse Math.* 30, 452–55. (248)

4   As their students see them, *Two-Year College Math. J.* 7, (2) 54, (3) 51, (4) 38.

5   Short dictionary of heuristic, (excerpts from *How to Solve It*), *A Taste of Science* (Ralph J. Tykodi, ed.), Westport, CT, Technomic, 1975, pp. 107–13.

1977

1* *Mathematical Methods in Science* (ed. Leon Bowden), New Mathematical Library, The Mathematical Association of America. (Compare [1963, 3].)

2   A note of welcome, *J. Graph Theory* 1, 5.

1979

1   More on guessing and proving, *Two-Year College Math. J.* 10, 255–58.

1982

1   On my cooperation with Gábor Szegő, *Gábor Szegő: Collected Papers*, vol. 1, Birkhäuser, p. 11.

1983

1* *Notes on Introductory Combinatorics* (with R. Tarjan and D. Woods), Birkhäuser. (Translation: Japanese, 1986.)

2   Mathematics promotes the mind, *Proc. Fourth International Congress on Mathematical Education*, Birkhäuser, p. 1.

1984

1   On problems with solutions attainable in more than one way (with Jean Pedersen), *College Math. J.* 15, 218–28.

2* *George Pólya: Collected Papers*. Vol. III. *Analysis* (ed. J. Hersch and G.-C. Rota). Vol. IV. *Probability, Combinatorics, Teaching and Learning Mathematics* (ed. G.-C. Rota), MIT Press. (See [1974, 3].)

1987

1* *Combinatorial Enumeration of Groups, Graphs, and Chemical Compounds* (with R. C. Read), Springer. (See [1937, 3].)

2* *The Pólya Picture Album: Encounters of a Mathematician* (with G. L. Alexanderson, ed.), Birkhäuser.

A list of problems posed and solved is unusual in a bibliography, but because of Pólya's special interest in and contributions to problem-solving, we list here problems and solutions he published in a variety of journals. Some of these have been often cited in the literature and led to a number of subsequent investigations.

P = problem   S = solution

Amer. Math. Monthly: 51 (1944), 96, P 4108; 51 (1944), 167, P 4111; 51 (1944), 533 P 4138; 51 (1944), 593, P 4142; 53 (1946), 279–82, S 4138; 53 (1946), 591, P E748; 54 (1947), 107, P E756; 54 (1947), 473, S E756; 54 (1947), 340, P E780; 54 (1947), 346, P 4255; 54 (1947), 479, P 4264; 55 (1948), 162, P E780.

Arch. Math. Phys., Ser. 3: 20 (1913), 271–72, P 424, P 425, P 426, P 427, P 428; 21 (1913), 181–85, S 383; 21 (1913), 288, 290, P 451, P 453, P454, S 398; 21 (1913), 366–68, S 400; 21 (1913), 370–71, S 427, S 428; 23 (1915), 289, P 486, P 487; 24 (1916), 84, P 498, P 499, P 500, P 501, P 502; 24 (1916), 282–83, P 509, P 510, P 511, P 512, P 513; 24 (1916), 369–75, S 386; 25 (1917), 85, P 520; 25 (1917), 337, P 535, P 536, P 537, P 538; 26 (1917), 65, P 542; 26 (1917), 66, S 500; 26 (1918), 161–62, P 561, P 562, P 563, P 564, P 565; 28 (1920), 173–74, P 584, P 585, P 586, P 587.

Elem. Math.: 16 (1961), 92, P 408.

Interméd. des Math.: 20 (1913), 57–58, S 3241; 20 (1913), 127–28, S 339; 20 (1913), 145–46, P 4240; 21 (1914), 27, P 4340.

Interméd. des Math., Sér. 2: 1 (1922), 81–82, S 2917; 1 (1922), 85–86, S 4240, S 4340; 4 (1925), 74–75, P 5517; 4 (1925), 82–83, S 5100.

Jahresber. Deutsch. Math.-Verein., Ser. 2: 32 (1923), 16, P 11; 34 (1925), 97–98, P 23, P 24, P 25; 35 (1926), 48, P 35, P 36, P 37; 37 (1928), 82–83, S 24; 40 (1931), 6, S 23; 40 (1931), 80, P 103 (with R. Nevanlinna); 40 (1931), 80–81, P 104, P 105, P 106, P 107, P 108; 43 (1933), 14–15, S 131; 43 (1933), 67–69, S 105.

Math. és. Term. Ért.: 32 (1914), 662-65.

Math. Mag.: 28 (1955), 235-36, S 209.

Nouvelles Annales Math., Ser. 4: 11 (1911), 377–81, S 1579; 11 (1911), 382, S 1580; 11 (1911), 383–84, S 1661.

Appendix 1

Pólya's Work in Probability

K. L. Chung

Pólya's first publication, his 1913 Dissertation [1912, 1] in Hungarian, treats a problem in probability. His lifelong interest in this field is evident from his Bibliography (see [1984, 2]). Besides the twenty papers reprinted in that volume, some ten more titles indicate probability and statistics. Some of his contributions to probability are classified as analysis, others perhaps as combinatorics. The major items which will be discussed here have long since become textbook material, [1]. Readers wishing to learn more details of the results summarized below should consult these texts.

1. Fourier transform and convergence of distributions

The Fourier transform f of a probability measure μ in R^1 given by

$$(1) \qquad f(t) = \int_{-\infty}^{\infty} e^{itx} \mu(dx), \quad t \in R^1$$

is known as a characteristic function, a tool used since Laplace, Poisson and Cauchy. In [1923, 2], Pólya showed that it is indeed characteristic in the sense that it uniquely determines μ. Actually he considered an absolutely Riemann-integrable function (signed density), and said of this result: 'despite its simplicity it was nowhere explicitly stressed.' So he gave a neat proof using Fejér's theorem in Fourier series. This result has been superseded by Paul Lévy's inversion formula but the method of proof remains viable. A better known result of Pólya's, slightly hidden in [1918, 4] ('On zeroes of certain integral functions') but cited and used in [1923, 2], is his sufficient condition for a given function to be a characteristic function, as follows:

Reprinted by permission from the *Bulletin of the London Mathematical Society* 19 (1987), 559–608.

(2) f is real-valued and continuous in R^1;

$f(0) = 1, f(t) = f(-t)$ for all t, f is convex for $t > 0$, $\lim_{t \to \infty} f(t) = 0$.

He returned to it in [1949, 3] and gave several interesting examples. This criterion is still the only useful one for constructing and recognizing specific characteristic functions, the 'positive-definite' characterization by Bochner and Khintchine being too perfect to be of practical utility. It should be pointed out that Pólya considered only probability density so that the last condition in (2) is necessary by the Riemann-Lebesgue lemma. For a general measure μ as written in (1) that last condition should be omitted.

Pólya wrote a number of papers on the 'Gaussian error-law', now commonly known as the 'normal distribution'. In [1920, 3] he coined the term 'central limit theorem' (in German) and proved a general convergence theorem for a sequence of distributions, based on the convergence of transforms. Let F be a distribution function and

(3) $$\Phi(u) = \int_{-\infty}^{\infty} e^{ux} dF(x)$$

be the transform. He said: Tschebyscheff and Markoff considered the derivatives of Φ at $u = 0$, whereas Liapounoff considered Φ for purely imaginary values of u. But he would consider Φ as an analytic function in the strip $-a \leq \operatorname{Re} u \leq +a$, $a > 0$. This requires the convergence of the integral in (3) in a neighborhood of the complex parameter u at the origin, which restricts the general applicability of the resulting convergence theorem. However, Pólya also proved an earlier version of Lévy's theorem (1922) at the end of [1926, 9]. Later in 1937 Lévy and Cramér published their convergence theorem in terms of characteristic functions (namely Liapounoff's usage) which is now the principal tool for limit theorems. Pólya also proved in [1920, 3] the so-called 'continuity theorem for the moment problem': namely if all moments of a sequence of distributions are finite and converge to those of a given distribution which is uniquely determined by its moments, then (weak) convergence of the sequence to the latter follows. This remains a useful and convenient method in situations where the calculation of moments is more expedient than that of the transforms. For instance I was able to obtain a central limit theorem for a stochastic iterative scheme concocted by Robbins and Monro in a medical problem.

In [1923, 2] Pólya made another approach to the Gaussian error-law by considering a functional equation suggested by the combination of errors in measurement. In terms of the characteristic function Φ of the error-law, it reduces to the following: for each $a > 0$ and $b > 0$, there exists $c > 0$ such that

$$\Phi(ct) = \Phi(at)\,\Phi(bt)$$

Such a law is called 'stable' by Lévy. Pólya proved that the only stable law having a finite second moment is Gaussian corresponding to $\Phi(t) = e^{-t^2}$. He deduced from this that the function $e^{-|t|^\alpha}$ cannot be a characteristic function if $\alpha > 2$, while it is a characteristic function if $0 < \alpha \le 1$, an immediate consequence of his condition (2). Lévy proved that the same is true for $0 < \alpha \le 2$ by using his convergence theorem, and discovered a much larger class of characteristic functions called 'infinitely divisible,' and even more profoundly, founded his theory of 'additive processes.' See [2].

2. Random walk

Pólya's celebrated theorem on random walks appeared in [1921, 7], but important complements are given in [1938, 2], based on a lecture he gave at a conference in Geneva in 1937, which was accompanied by a film showing 'the cartage of stones by the current.' A point (not a 'particle'!) executes a walk on the integer lattice of R^d, $d \ge 1$, so each step takes it to one of the $2d$ neighboring positions with probability $1/2d$ each. The steps are independent. Given an arbitrary lattice point A, will the point starting from the origin ever reach A with probability one? The answer is 'Yes' if $d = 1$ or 2, but 'No' if $d \ge 3$. He stressed the dimensional breakdown as 'newsworthy,' not intuitively obvious. For $d = 1$ the problem is similar to that of gambler's ruin treated by Laplace. But the question raised by Pólya, that of *recurrence* as it is called today, lay hidden in the ruin problem. One might say that the demand of a quantitative answer, the probability of ruin of Peter before Paul, obscured the even more fundamental question of the certainty of ruin (of one of them). This question becomes more prominent in higher dimensions and opens up a new vista. It can also be formulated as the problem of *rencontre* (Pólya reminded the reader that he was not speaking of the classical problem of the Marquis de Montmort!) of two points executing random walks independently. He treated this second problem by analogy with the first, and added a third one in [1938, 2], that of 'novelty': will the point never pass through the same position twice? He reduced this to the first problem by an elegant 'reversal' argument. Actually the second problem can also be reduced to the first by reversing the steps of one of the strollers from the site of rencontre. This kind of reasoning has received some attention lately under the name of 'coupling.' Pólya proved his results by using a 'first passage decomposition'. When $A = 0$, if we denote the position of the point at time n by S_n, and put $p_0 = 1$,

$$p_n = P(S_n = 0), \ w_n = P(S_k \neq 0 \text{ for } 1 \leq k < n; \ S_n = 0), \ n \geq 1;$$

$$P(z) = \sum_{n=0}^{\infty} p_n z^n; \quad W(z) = \sum_{n=1}^{\infty} w_n z^n.$$

Then he obtains the relation (known today as 'renewal equation'):

(4)
$$P(z) = \frac{1}{1 - W(z)}.$$

From this he easily deduces that

(5)
$$\sum_{n=1}^{\infty} w_n = 1 \Leftrightarrow \sum_{n=0}^{\infty} p_n = \infty.$$

Now he calculates p_n by Fourier inversion and Laplace's method (apparently used in his dissertation). The result is

(6)
$$p_n \sim 2 \left(\frac{d}{2\pi} \right)^{d/2} n^{-d/2},$$

from which it follows that the conclusion in (5) obtains if and only if $d = 1$ or 2. The general case $A \neq 0$ follows from this special case and another relation of the form $P_A(z) = W(z) P(z)$, with the same P as before.

It is clear in Pólya's walk that if return to 0 once is (almost) certain then so also is return infinitely many times. We now call such a walk 'recurrent.' Equally clear is the other side of the dichotomy: if return once is uncertain then return infinitely many times is impossible. We call such a walk 'transient.' Thus, 'non-recurrent' has a stronger implication than it logically connotes. The change of viewpoint from 'at least once' to 'infinitely often' might seem specious. It is decidedly not; indeed the switch is tantamount to converting a supermartingale to a martingale, or a superharmonic function to a harmonic one. In the present case it yields the following: if N denotes the total number of returns to 0, and $E(N)$ its mathematical expectation, then

(7) $P(N = \infty) = 0$ or 1 according as $E(N) < \infty$ or $= \infty$.

In this form the theorem constitutes an instance of the Borel-Cantelli lemma, but since the events (returns) are not independent the result is not covered by the original assertion. In this light Pólya's theorem appeared to be the first significant case of a 'zero-to-one' law for dependent events, an outstanding phenomenon in probability.

An extension of Pólya's theorem to 'generalized random walks' was obtained by Chung and Fuchs [3]. Let $\{X_n, \ n = 1, 2, \ldots\}$ be a sequence of independent and identically distributed random variables in R^d, and $S_n = \sum_{k=1}^{n} X_k$. Thus in Pólya's scheme the X_k denote the successive steps, and S_n the position at time n, the walk starting from 0. A punctilious return to 0 is now ruled out in general, and what can be more natural than to substitute for this a return to an arbitrary neighborhood B of 0? Note that when the latter is a ball of radius smaller than 1, then in the lattice case return to the ball means return to 0 precisely. It is rather unexpected that the dichotomy presented in (7) remains true if N now denotes the total number of visits to B by the sequence S_n, $n = 1, 2, \ldots$. To the unspoiled it may also be surprising that the dichotomy does not depend on the size of B. To evaluate the sum represented by $E(N)$ there, Fourier inversion with Abel summability was used. The conclusions are: if $d = 1$, the random walk is recurrent if $E(X_1) = 0$; if $d = 2$, it is recurrent under this condition and the supplementary condition $E(||X_1||^2) < \infty$; if $d \geq 3$, it is always transient (unless degenerate into a lower dimension). Further results, along this direction were given by Kesten, Spitzer, Port, Stone, Dudley,..., some of which deal with walks on groups. Extension of Pólya's theorem in the form (7) to homogeneous Markov chains with countably many states is quite easy. In fact, the method indicated above generalizes easily with suitable notation, which would have rid Pólya of the nuisance of 'periodicity' pestering him: return is possible only in an even number of steps. The general notion of recurrence vs. transience in Markov processes in discrete or continuous time plays a prime role in the theory. It has been investigated by many probabilists: Doblin, Harris, Orey, Azéma, Kaplan-Duflo, Revuz.... The case of Brownian motion in R^d deserves special mention. In [1938, 2] Pólya discussed the diffusion of molecules as well as Einstein's study of cartage. He wrote down the heat equation, a hyperbolic equation (for H. A. Einstein's cartage problem) as well as the Laplace equation, by passage to the limit of second order difference equations. Thereby he indicated a heuristic solution to his old problem by considering positive harmonic functions, pointing out in particular the solution $1/r$ (where r denotes the distance from the origin) conveys the transience in R^3. Indeed in the last footnote to the paper he discussed the probability interpretation of a Dirichlet boundary value problem with values zero or one on different parts of the boundary. This is essentially the modern method used to ascertain the recurrence or transience of Brownian motion in different dimensions. Let us note that the results on generalized random walks stated above imply the recurrence, but not the transience cases. Pólya considered the application of probability to the 'transport of solid material by the rivers' as 'serious', whereas his own scheme of 'promenade au hazard' as 'curious.' In [1970, 2] he told the story of how he came to conceive of the latter from constantly running into a couple while strolling in the park. Actually, shortly before he wrote [1919, 9] he had treated Lord Rayleigh's problem of random flight in [1921, 7], and used the same word 'Irrfahrt' in both papers. That earlier, no doubt serious, application apparently did not meet Descartes' precept, which Pólya quoted, of 'commencing with the objects that are the simplest and easiest to know,' and it yielded no memorable consequences. On the other hand, half a century since Pólya wrote those charming passages, the curious has certainly become curiouser and curiouser.

3. Urn scheme

This is also known as the Pólya-Eggenberger scheme (see [1923, 7] and [1928, 6]). Eggenberger wrote his thesis in Zürich under Pólya in 1924, and did the practical work. The best exposition of the ideas, however, was given in [1931, 1] which reproduced the contents of a course Pólya gave in March 1929 at the Institut Henri Poincaré, for which he thanked Émile Borel for the invitation. Pólya was seeking non-Gaussian laws of errors for dependent events, and cited Markov's chained-events. The spread of epidemic was mentioned as an example, and led to the following model of 'contagion.' An urn contains initially ρ red and σ black balls, and one ball is drawn each time. After each drawing, $1 + \delta$ balls of the color drawn are added to the urn, where δ is an integer and adding becomes subtracting when the number is negative. Call the drawing of a red ball 'success'. If $\delta > 0$ then success is contagious as it reinforces other successes. If $\delta = 0$ or -1, the scheme reduces to the classical ones of drawing with or without replacement, which led respectively to the central limit theorem (Bernoulli, de Moivre, Laplace, Gauss) and 'the law of small numbers' (Poisson). So Pólya's extension is certainly very 'simple'; but is it also very 'easy to know' according to Descartes' precept? Let us ask the most exigent question about the scheme: what is the probability of drawing n balls with their colors completely specified in sequential order? An easy argument gives the answer

$$(8) \quad \frac{\rho(\rho+\delta)(\rho+2\delta)\cdots(\rho+(r-1)\delta)\sigma(\sigma+\delta)(\sigma+2\delta)\cdots((\sigma+(s-1)\delta)}{1(1+\delta)(1+2\delta)\cdots(1+(r-1)\delta)(1+r\delta)(1+(r+1)\delta)\cdots(1+(n-1)\delta)},$$

where r denotes the total number of red balls in the specified sequence, s that of the black ones, so that $r + s = n$. Given the initial data σ and ρ, this probability depends only on n and r, but not on the specified order in which the two colors appear, so long as none of the factors in (8) becomes negative. Thus the formula holds for all $n \geq 1$ when $\delta \geq 0$. A sequence of random variables $x_1, x_2, \ldots, x_n, \ldots$ is now called 'exchangeable' when the joint distribution of any n (arbitrary) of them in any permutation is the same. If we put x_n equal to one or zero according as the nth ball drawn in Pólya's scheme is red or black, we obtain such a sequence. This is a remarkable generalization of a sequence of independent and identically distributed random variables that constituted the primary structure of classical investigation. The theory of exchangeability has been developed by Bruno DeFinetti with implications on the foundation of statistics [4]. The concept implies that of (strict) stationarity, but even for $n = 1$ it is not trivial to show that the probability of drawing a red ball at the nth drawing is the same for all n. In case $\delta = -1$, Poisson had to give a combinatorial argument that was not so easy to follow. Pólya gave an elegant proof of stationarity in the general case by use of a multiple generating function. Clearly, here probability merges into combinatorics.

Pólya's distribution given by (8) can be reorganized by using generalized binomial coefficients into a hypergeometric form, now known by his name. If we let the parameters ρ, σ, and δ grow (or not grow) with n and let n become infinite, we obtain various limiting distributions including the normal, Poisson (compounded), gamma, beta, etc. (but not all Pearson types). Applications to special processes have been made in various areas of practice, too numerous to recount here. For a curious and serious application, read the title of [5]. This brief discussion must not stop before a note of dissonance, sounded by Pólya himself. He noted that if $\delta > 0$ in his scheme, failure as well as success tends to reinforce itself. Now in an epidemic each victim produces many more germs to cause further infection, but why should each uninfected person also enhance the chance of others being so? Here we must hear his own voice (translated, italics mine): 'In reducing this fact [of contagion] to its simplest terms and adding to it a certain symmetry, *propitious for mathematical treatment,* we are led to the urn scheme.' Judging from the data shown in [1923, 7], theory and practice fit well enough in the case of Swiss smallpox.

4. Miscellanies

A number of papers reprinted in other volumes of the *Collected Papers* bear fruits in probability. [1932, 2] (with A. Bloch) deals with roots of equations with random $(0, \pm 1)$ coefficients. Since Pólya spent a preponderant time studying the roots of functions (originally as an approach to the Riemann Hypothesis), it is fitting that he should randomize the problem at least once. This topic was taken up by Littlewood and Offord later. [1931, 4] (with Szegő) dealing with the transfinite diameter was a sequel to his earlier work [1928, 5]. It played an important role in the development of the notion of 'capacity,' through M. Riesz, Frostman and Choquet. The paper was roundly commented on by Hille. Since G. A. Hunt, P. A. Meyer and Dellacherie, capacity has become a germane part of probability. It is possible that we can still learn something from the old source. [1933, 3] dealing with heat exchange inspired Schoenberg and Karlin to the study of so-called Pólya frequency functions with statistical applications [6].

Among the remaining reprinted papers two early ones belong to geometrical probability and have been adequately commented on by Santaló, Coxeter and Kingman. A few others solve special problems such as balloting, coupon-collecting and clustering of randomly picked points. One is on heuristic reasoning but another [1921, 1], which gives an intuitive derivation of the Gaussian law, was not reprinted. Pólya presented the latter in my class some years ago, together with a new statistical-methodological characterization of that law, which he announced at the Bologna Congress [1928, 8] and later discussed in detail in [1931, 1], cited in Section 3 above. I recall that he asked in class whether his assumptions for solving the problem could not be ameliorated. The last item in probability which is reprinted [1976, 1] contains a simple problem with an empirical twist, highly recommended to teachers of elementary courses.

References

1. See, for example, both volumes of W. Feller's *An introduction to probability theory and its applications* (John Wiley & Sons, 1950, 1966), or my *A course in probability theory* (2nd ed., Academic Press, 1974) and *Elementary probability theory with stochastic processes* (3rd ed., Springer, Berlin, 1979).
2. P. LÉVY, *Théorie de l'addition des variables aléatoires* Gauthier-Villars [first edition published in 1937]).
3. K. L. CHUNG and W. H. J. FUCHS, 'On the distribution of values of sums of random variables', *Mem. Amer. Math. Soc.* 6 (1951).
4. An early exposition was given at the same conference in Geneva as Pólya did in [1938, 2]; see *Actualités scientifiques et industrielles*, 739 (Hermann & Cie.).
5. D. BLACKWELL and D. G. KENDALL, 'The Martin boundary for Pólya's urn scheme, and an application to stochastic population growth', *J. Appl. Probability* 1 (1964) 284–296.
6. See for example, S. KARLIN, *Total positivity* (Stanford University Press, 1968).

Appendix 2

Pólya's Work in Analysis

R. P. Boas

When one looks at Pólya's work as a whole one is struck by his great power and versatility. He proved many difficult, fascinating and important theorems, and devised methods that still retain their effectiveness today. It is interesting to see how, in many cases, he found his problems. Some of them seem to spring up from nowhere, but he often makes use of the principle (enunciated in the preface to *Aufgaben und Lehrsätze*) that to determine a line one can start at a point and follow a direction, or interpolate between two points, or draw a parallel. The difference between Pólya and other people was that his generalizations were usually both deep and difficult; or at least, unexpected. He was also quick to see where something interesting was going on, and step in with an improvement, or even a complete theory. Hardy is supposed to have said once that Pólya had brilliant ideas but didn't follow them up. There was some truth in this unkind remark. The *Collected Papers* include many brief contributions that contain the germs of substantial theories that were developed later by others. Nevertheless, it would be unreasonable to complain considering that at the height of his career Pólya was publishing two or three major papers in analysis every year and doing the same thing in probability. He could hardly have had time to follow up everything that he initiated. In parts of analysis his best ideas were tossed out as problems (many of which led to substantial theories), or absorbed into *Aufgaben und Lehrsätze* or into *Inequalities*. He was also very helpful to others. I can illustrate this trait from my own experience. In the early 1940s I was interested in what is

Reprinted by permission from the *Bulletin of the London Mathematical Society* 19 (1987), 559–608.

now known as the Whittaker constant W. It was known that if f is an integral function of exponential type τ there is a number W such that if f and all its derivatives have zeros in the unit disk, then $f \equiv 0$ if $\tau < W$ but not necessarily if $\tau > W$; and that $\log 2 \leq W \leq \pi/4$. I wrote to Pólya about some ideas I had about this problem, and he replied with a letter that improved both upper and lower bounds, with suggestions for how to make further improvements; but he never published anything on the subject himself.

In discussing Pólya's work in Analysis, it seems best to begin with one of the fields to which he made the most numerous and most significant contributions, namely complex analysis.

1. *Connections between the sequence of coefficients in a power series and properties of the analytic function determined by the power series.* In principle, the sequence of coefficients contains all the properties of the function; the problem is to make the sequence surrender information about some particular property. The most interesting theorems connect an easily-stated property of the coefficients with an easily-stated property of the function. Pólya's own survey of this field is in [1928, 7].

Fatou conjectured, and Pólya proved [1916, 1; 1929, 1] that the circle of convergence is 'usually' a natural boundary: by changing the signs of the coefficients one can always make the sum of the series non-continuable. In fact [1917, 7], this happens for almost all sequences of signs in the sense that the set of sequences for which it happens is the complement of a nowhere dense perfect set. This must have been one of the first applications of point-set topology to complex analysis. Pólya also showed [1950, 4] that it is possible to change the signs so that the sum of the series satisfies no algebraic differential equation.

Another easily-stated property of the sequence of coefficients is the presence of many gaps. According to Fabry's famous gap theorem, a series is non-continuable if the density of zero coefficients is 1. Pólya established the definitive character of Fabry's theorem by showing [1939, 1; 1942, 1] that no weaker hypothesis than Fabry's will allow the same conclusion. If the density of zero coefficients is less than 1 but still positive, Pólya showed [1923, 4; 1929, 1] that singular points still occur, but less frequently; in particular, if the minimum density of the vanishing coefficients is δ, there is no arc of regularity whose length exceeds $1-\delta$ times the circumference. (This had been proved by Fabry, but in a less lucid formulation.) In [1923, 5] Pólya proved a much deeper result, the corresponding theorem for Dirichlet series on the line of convergence.

Pólya also extended Fabry's theorem in another way, by considering Dirichlet series whose exponents are not necessarily real [1927, 10, 11]: here, if $n/\lambda_n \to 0$, the domain of the function is necessarily convex; Fabry's theorem is an immediate consequence. This is one of the few known significant results about Dirichlet series with complex exponents. For the special case of power series with $\liminf n/\lambda_n = 0$, Pólya showed that the domain is simply-connected. In [1934, 3] Pólya extended these investigations to functions that do not have single-valued continuations. The results are rather complicated; the following special case gives an idea of their character. A function defined by a power series $\sum a_n z^n$ is of the form $F\{-\log(1-z)\}$, where F is an integral function at most of order 1 and mean type, if and only if $a_n = H(n)$, where

$$H(z) = \{1 / \Gamma(z+1)\} \sum_{k=0}^{\infty} A_k \Gamma^{(k)}(z) / k!,$$

with $\limsup |A_n|^{1/n} < \infty$.

Instead of considering zeros in the sequence of coefficients, one can connect the changes of sign in the sequence with the location and nature of the singular points. The idea appears in its purest form in [1932, 1], where the sum of the power series has poles, and no other singularities, on the unit circle; the spacing of the poles is connected with the variations in sign of the sequence of coefficients. This, together with results about trigonometric polynomials and about Dirichlet series, follows from a theorem about Mellin transforms proved in [1930, 3]; it takes someone of Pólya's skill and breadth of knowledge to perceive the connections.

Many relations between the coefficients and the singular points of a power series are expressed in terms of recurrent determinants formed from the coefficients. Paper [1928, 4] provides a condition of this kind for the sum of the series to have at least one regular point on the boundary.

Still different results on singular points are connected with the Hadamard product $\sum a_n b_n z^n$ of $A(z) = \sum a_n z^n$ and $B(z) = \sum b_n z^n$. The singular points of the product can in general occur only at products of singular points of A and B; Pólya devoted three papers [1927, 2, 3; 1933, 5] to the question of when a product of singular points will necessarily be a singular point of the product.

The coefficients of many familiar power series are integers. The question of the behavior of the sum of such a series is, as Pólya remarked in his survey [1922, 3], an attractive one, because it combines the apparently unrelated concepts of rational functions, integers, and conformal mapping. Pólya conjectured [1916, 3] and Carlson subsequently proved [1], that a power series with integral coefficients and radius of convergence 1 represents either a rational function or a function with the unit circle as a natural boundary. In [1923, 1] and [1928, 4] Pólya returned to this subject and found still deeper results.

The methods that Pólya used in proving many of the results mentioned so far were systematized in a long paper [1929, 1], which Pólya himself partially summarized in [1929, 3]. This very influential paper studies various kinds of densities of sequences of numbers; convex sets and their supporting functions; and integral functions of exponential type and their conjugate indicator diagrams (which describe the rate of growth of functions in different directions). Finally this material is applied to theorems about the distribution of values of integral functions of infinite order, and the location of singular points of functions whose power series have a finite radius of convergence. All these topics have turned out to be capable of generalizations and extensions that go far beyond those actually presented in the paper. The central theorem is Pólya's representation of an integral function f of exponential type (one whose modulus grows no faster than $\exp(A|z|)$) as

$$\frac{1}{2\pi i}\int_C F(w)e^{zw}\,dw$$

where F is the Borel-Laplace transform of f, and C surrounds the conjugate indicator diagram. This representation has turned out to be important in contexts far beyond those to which Pólya originally applied it. For example, expansion of the kernel e^{zw} in series leads to expansions of functions of exponential type in terms of various data. Pólya applied this idea on a small scale in [1926, 1]; the same idea was applied to more substantial problems by his students and by many others.

A related paper with Plancherel [1937, 2] was ostensibly concerned with generalizations to n dimensions of the Paley-Wiener theorem (a function f is of the form

$$\int_{-\tau}^{\tau} e^{izt}g(t)\,dt,\ \ g \in L^2$$

if and only if f is an integral function of exponential type at most τ and belongs to $L^2(-\infty, \infty)$); however, the same paper contains many important results on one-dimensional problems.

The last chapter of [1929, 1] appears as [1933, 5], which contains a full discussion of Pólya's results on the singular points of an Hadamard product.

Fabry's gap theorem and its generalizations do not, of course, apply to integral functions, for which there is only one singular point. What, then, does a gap condition for the power series of an integral function imply? Pólya answered this question in [1928, 4], although the proofs appear only in [1929, 1]: for each theorem about singular points on the circle of convergence of a power series there is a parallel theorem on Julia lines of integral functions.

The theorems on power series with integral coefficients were extended in [1931, 2] to power series whose coefficients are integers of an algebraic field. Related problems concern power series $\sum a_n z^n$ with only finitely many different a_n [1931, 2] and those for which $n! a_n$ are integers [1921, 6; 1922, 4]. In [1922, 2] Pólya studied power series whose sums are algebraic functions (and shows how to construct such series). In [1935, 3] and [1950, 4] Pólya studied power series whose sums satisfy an algebraic differential equation. In particular, he proved that a formal power-series solution must actually converge. Paper [1920, 1] discusses rational functions whose power series have rational coefficients.

2. *The general character of an analytic function as revealed by its behavior on a set of isolated points.* The subject originates with Pólya's discovery [1915, 2] that 2^z is the 'smallest' (in a well-defined sense) transcendental integral function with integral values at the positive integers. Pólya's further contributions to this topic are in [1916, 3] and [1920, 4; 1928, 4]; these results have inspired much further work, which peaked several decades later. The function 2^z is (in the terminology introduced later by Pólya) of exponential type $\log 2$. For somewhat larger type, an integral function with integral values at the positive integers must be of the form $P(z)2^z + Q(z)$, where P and Q are polynomials; this was discovered by A. Selberg [2]. Activity in this field continued at least into the 1970s (for a survey, see Vol. 1 of the *Collected Papers*, pp. 771–772).

Paper [1941, 3] is a fundamental contribution to the interpolatory theory of integral functions: what kinds of function can take, with some of their derivatives, prescribed values at prescribed points? Papers [1933, 6] and [1937, 2] deal with the problem of how slow the growth of an integral function can be when it is bounded on a regularly spaced sequence of points. The problem that inspired [1933, 6] was a conjecture by Littlewood that an integral function of order less than 2 cannot be bounded at the lattice points unless it is a constant. This had been proved by J. M. Whittaker [3], but Pólya's method is at the basis of many subsequent generalizations. Paper [1933, 6] is presented as a commentary on a problem set by Pólya in 1931: an integral function of order 1, type 0, bounded at the integers, is constant (this had been proved earlier by Valiron [4], although Pólya was not aware of this). This again has led to an extensive array of generalizations.

3. *General theory of analytic functions.* Paper [1926, 3] contains the first proof of the theorem that if g and h are integral functions and $g(h(x))$ is of finite order, then either h is a polynomial and g is of finite order, or g is of zero order and h is of finite order. Paper [1926, 4] contains an elegant proof of the 'cos $\pi\rho$' theorem (originally conjectured by Littlewood [5] and proved by Valiron [6] and Wiman [7]): if f is of order ρ, type 0, where $0 < \rho < 1$, or of smaller order, then

$$\limsup [\log m(r)]/[\log M(r)] \geq \cos \pi\rho,$$

where m and M are the minimum and maximum moduli of f.

Papers [1927, 4] and [1928, 2], with Hardy and Ingham, are concerned primarily with theorems of Phragmén-Lindelöf type for functions that are analytic (or subharmonic) in a strip. The general question is, what can be deduced about the growth of f from the growth of

$$\phi(x, y) = \frac{1}{2y} \int_{-y}^{y} |f(x + iw)|^p \, dw$$

in a strip $\alpha \leq x \leq \beta$ (or its closure). Some of the results are surprising, and most of the proofs are difficult.

4. *Zeros of polynomials and other analytic functions.* Pólya was particularly fond of theorems that connect properties of an integral function with properties of the sets of zeros of polynomials that approximate the integral function. He dealt with this general problem in [1913, 8, 9; 1914, 3]; a great deal of later work starts from there. In particular, in these papers he showed that all multiplier sequences that transform polynomials with real zeros into polynomials with real zeros are generated by the special entire functions of types I and II in [1914, 5] and [1915, 1]. These functions are defined as follows:

(I) $$\Phi(x) = \frac{\alpha_r}{r!} x^r e^{\gamma x} \prod_{\nu=1}^{\infty} (1 + \gamma_\nu x), \quad \gamma \geq 0, \ \gamma_\nu \geq 0;$$

(II) $$\Psi(x) = \frac{\beta_r}{r!} x^r e^{-\gamma x^2 + \delta x} \prod_{\nu=1}^{\infty} (1 + \delta_\nu x) e^{-\delta_\nu x}, \quad \delta_\nu \text{ real}, \ \gamma \geq 0$$

(the zero function is considered to belong to both classes). These functions are now often referred to as Pólya-Schur or Laguerre-Pólya functions. Functions of class I are characterized as limits of polynomials with only real zeros, all of the same sign; functions of class II, as limits of polynomials with only real zeros. A power series $\sum_{n=0}^{\infty}(\gamma_n / n!)x^n$ belongs to class II or I according as the polynomial $\sum_{k=0}^{n} x^{n-k}\gamma_k\binom{n}{k}$ has, for all n, either all its zeros real or all its zeros real with the same sign.

There are now many applications of the Laguerre-Pólya functions: they underlie the general inversion theory of convolution transforms (Hirschman and Widder [8]); the theory of variation-diminishing transforms [9], interpolation by spline functions [10], and many other applications. Some more direct applications are given in [1914, 6; 1915, 4; 1927, 6]; these papers require separate notice because they have to do with two other themes that occur frequently in Pólya's work.

Paper [1927, 6] is one of a series dealing with zeros of trigonometric integrals; as Pólya explained in [1918, 4] and [1927, 6], his motivation for this work was that the Riemann ξ-function is represented by a trigonometric integral, so that a sufficiently good theorem about the zeros of trigonometric integrals would establish the Riemann hypothesis. That this hope is almost certainly illusory hardly diminishes the interest of the theorems that Pólya found. In [1918, 4] he started with the Fourier transform of a function supported on [−1, 1], and showed that in certain cases it has only real zeros. Paper [1920, 5] presents almost everything that one would want to know about zeros of exponential polynomials, but without proofs; the proofs are available only in a Zürich thesis by Pólya's student E. Schwengeler [11]. This work led into the modern theory of the zeros of integral functions of exponential type, and also into [1933, 6] and [1937, 2], although Pólya did not contribute very much directly to this subject. In the main, Pólya concentrated on trigonometric integrals over $(-\infty, \infty)$, identifying progressively more general classes of integrals that have only real zeros [1918, 4; 1923, 8; 1926, 5, 8; 1927, 6, 7]. Although some functions that closely resemble the ξ-function do have only real zeros, they do not do much for the Riemann hypothesis; but these papers and their generalizations have been useful for other purposes: notably in physics, as Kac brings out in his comments on [1926, 8] in Vol. 2 of the *Collected Papers*.

In [1927, 7] Pólya raised the question of whether a certain family of inequalities (now known as the Turán inequalities) are satisfied; since they form a necessary condition for the truth of the Riemann hypothesis, that hypothesis would be disproved if any one failed. The first progress on this question was made in 1966 when the inequalities were proved for a sufficiently large index [12]; the question was finally settled in 1986 [13].

Pólya devoted a great deal of attention to the question of how the behavior in the large of an integral or meromorphic function influences the distribution of the zeros of successive derivatives. His principal papers on this topic are [1914, 6; 1915, 4; 1921, 4; 1937, 4]; the survey [1943, 1] covers almost everything that was known up to 1942. He introduced the term 'final set' of an integral or meromorphic function for the set of limit points of the set of zeros of successive derivatives (counting also the points that are zeros of infinitely many derivatives). In [1922, 1] he determined the final sets of meromorphic functions: the final set is the polygon whose points are equidistant from the two nearest poles. It is much more difficult to determine the final sets of integral functions. Some results up to the early 1970s are mentioned in the comments on [1943, 1] in Vol. 2 of the *Collected Papers;* since then there has been considerable progress. For example, Pólya had showed [1937, 4] that for an integral function that is real on the real axis, and has only finitely many non-real zeros, the final set is a subset of the real axis provided that the order is less than 4/3; his conjecture that 4/3 can be replaced by 2 was established very recently [14]. Pólya remarked [1921, 4; 1922, 1] that when f is an integral function, real on the real axis, and f, f', and f'', have no zeros, then f is an exponential function. This has been generalized in various ways; in [13] it was shown that if f, f', f'', and f''' have only real zeros then either f is an exponential, or of the form $A(e^{icz} - e^{id})$ (c, d real), or a Pólya-Schur function; [15] also contains further results for meromorphic functions. It appears plausible that a Pólya-Schur function, of order greater than 1, real on the real axis, has the whole real axis as its final set; under some additional restrictions, this was established in [16].

In [1913, 3] Pólya gave the first correct proof of Laguerre's famous theorem, which states (loosely) that if f is the Laplace transform of $\phi, x > x_0$, and ϕ has V changes of sign, then f has at most V zeros on $x > x_0$. More precise results were found some 20 years later [17]. This is only one of many 'sign rules' for zeros; in particular, Sylvester's rule [1914, 1; Theorem IV] remained an isolated curiosity and was only explained in [1958, 3], 44 years later.

Paper [1916, 6] is an 'omnibus' theorem on the reality of zeros of a polynomial. It is worth quoting in full. Let $f(x) = \sum_{k=0}^{n} a_k x^k$ and $g(x) = \sum_{k=0}^{n+m} b_k x^k$ have only real zeros, $b_0, b_1, \ldots, b_n \geq 0$. Then the curve

$b_0 f(y) + b_1 x f'(y) + \cdots + b_n x^n f^{(n)}(y) = 0$ has n real intersections with every line $sx - ty + u = 0$, where $s \geq 0, t \geq 0, s + t > 0, u$ real. The special cases $x = 1, y = 0$, and $x = y$ are well known.

Paper [1932, 2] is noteworthy as having been the first paper on the zeros of a 'random' polynomial.

5. *Signs of derivatives and analytic behavior.* S. Bernstein [18] was the first to observe that a C^∞ function on a real interval, with all its derivatives positive, is real-analytic there; he also showed that the same conclusion follows if a sufficiently dense sequence of derivatives are positive. Much later, in 1940, D. V. Widder [19] discovered that if the derivatives of even order alternate in sign then the function is the restriction of an integral function of exponential type. This discovery inspired a substantial amount of activity, in which Pólya participated [1941, 1, 2; 1942, 2, 3] and which he surveyed in detail in [1943, 1]. The topic is still alive; see the comments on [1942, 2] and [1943, 1] in Vol. 2 of the *Collected Papers*; also [20].

6. *Conformal mapping.* Most of Pólya's work in this field is connected with the notion of transfinite diameter. In [1928, 4] this is connected with the recurrent determinants of the coefficients, and applied there and in [1928, 5] to Koebe's ¼ theorem, to properties of the coefficients adjacent to Hadamard gaps, and the continuability of power series. Pólya reformulated the usual geometric statements about univalent functions, and then generalized them to maps of multiply-connected regions. In [1931, 4], Pólya and Szegő extended the concept of transfinite diameter to three dimensions, thereby opening up a new field of research; they made a conjecture that was proved by Pólya and Schiffer: the transfinite diameter of a convex curve is no less that one-eighth of the perimeter [1959, 2]. In [1958, 3] Pólya and Schoenberg made a major contribution to mappings onto convex domains by univalent functions. This seminal paper inspired a great deal of further research.

7. *Real analysis.* It is arguable that Pólya's most important contributions to this area are in his share of the book [1934, 4] *Inequalities* by Hardy, Littlewood and Pólya. This was the first systematic study of the inequalities that are used by every working analyst. Although there are more recent books that contain more material in certain directions, this one is still a fundamental reference — even though Hardy was once heard to complain that whenever he needed an inequality, the precise one that

he wanted was not there. Pólya published little about inequalities himself, but did propose many problems about inequalities and series. Paper [1923, 3] is about the structure of real sequences; it is here that Pólya introduced what are now called Pólya peaks, which form an essential tool in many problems about integral functions. Paper [1950, 1] contains an inequality that has many applications to eigenvalues of operators.

Paper [1913, 5] is of interest as the first construction of a Peano curve with at most triple points (this being the smallest possible number).

Pólya was much interested in mean value theorems, whether for isolated functions or for solutions of differential equations. Paper [1921, 5] is a little-known gem in three dimensions. The theorems in [1922, 3, 4,] on mean value theorems corresponding to a linear homogeneous differential equation are still used extensively. In [1931, 5] Plancherel and Pólya studied the mean value

$$\lim_{R\to\infty}\int_{x-R}^{x+R} f(u)\,du = \phi(x).$$

They showed that if $\phi(x)$ exists for all x, it is necessarily a linear function. They also discussed a similar problem in two dimensions, where the limit is necessarily harmonic. Paper [1934, 1] deals with analogues of Rolle's theorem for partial differential operators.

Papers [1926, 7; 1927, 3; 1938, 1, 4] deal with moment sequences and the total indeterminacy of the Hamburger moment problem for functions of bounded variation, as well as with infinite systems of linear equations.

8. *Approximation theory and numerical analysis.* Papers [1914, 4; 1968, 1] deal with Graeffe's method for solving polynomial equations approximately. Until recently the method was not highly regarded because of the large amount of computation it requires, but with modern high-speed machines it is becoming useful again. Paper [1933, 2] was a pioneering investigation on numerical quadratures and is still an important result.

References

1. F. CARLSON, 'Ueber Potenzreihen mit ganzzahligen Koeffizienten', *Math. Z.* 9 (1921) 1–13.
2. A. SELBERG, 'Über ganzwertige ganze transzendente Funktionen', *Arch. Mat. Naturvid.* 44 (1941) 45–52.
3. J. M WHITTAKER, 'On the "flat" regions of integral functions of finite order', *Proc. Edinburgh Math. Soc.* 2 (1930) 111–128; *Interpolatory Function Theory* (Cambridge University Press, 1935), Chapter V.

4. G. VALIRON, 'Sur la formule d'interpolation de Lagrange', *Bull. Sci. Math.* (2) 49 (1925) 181–192, 203–224.
5. J. E. LITTLEWOOD, 'A general theorem on integral functions of finite order', *Proc. London Math. Soc.* (2) 6 (1908) 189–204.
6. G. VALIRON, 'Sur les fonctions entières d'ordre fini et d'ordre nul, et en particulier les fonctions à correspondence régulière', *Ann. Fac. Sci. Univ. Toulouse* (3) 5 (1914) 117–257.
7. A. WIMAN, 'Über eine Eigenschaft der ganzen Funktionen von der Höhe Null', *Math. Ann.* 76 (1915) 197–211.
8. I. HIRSCHMAN and D. V. WIDDER, *The convolution transform* (Princeton University Press, 1955).
9. I. J. SCHOENBERG, 'Über variationsvermindernde lineare Transformationen', *Math. Z.* 32 (1930), 321–328.
10. I. J. SCHOENBERG and A. WHITNEY, 'On Pólya frequency functions, III. The positivity of translation determinants with an application to the interpolation problem by spline curves', *Trans. Amer. Math. Soc.* 74 (1953) 246–259.
11. E. SCHWENGELER, *Geometrisches über die Verteilung der Nullstellen spezieller ganzer Funktionen (Exponentialsummen)* (Zürich, 1925).
12. E. GROSSWALD, 'Generalization of a formula of Hayman, and its applications to the study of Riemann's zeta function', *Illinois J. Math.* 10 (1966) 9–23; 'Correction and completion of the paper "Generalization of a formula of Hayman"' *ibid.* 13 (1969) 276–280.
13. G. CSORDAS, T. S. NORFOLK and R. S. VARGA, 'The Riemann hypothesis and the Turán inequalities', *Trans. Amer. Math. Soc.* 296 (1986) 521–541.
14. T. CRAVEN, G. CSORDAS and W. SMITH, 'The zeros of derivatives of entire functions, and the Pólya-Wiman conjecture', *Ann. of Math.* 125 (1987) 405–431.
15. S. HELLERSTEIN, L. C. SHEN and J. WILLIAMSON, 'Reality of the zeros of an entire function and its derivatives', *Trans. Amer. Math. Soc.* 275 (1983) 319–331.
16. LI-CHIEN SHEN, 'On the zeros of successive derivatives of even Laguerre-Pólya functions,' *Trans. Amer. Math. Soc.* 298 (1986) 643–652.
17. D. V. WIDDER, 'The inversion of the Laplace integral and the related moment problem', *Trans. Amer. Math. Soc.* 36 (1934) 107–200 (154–163).
18. S. BERNSTEIN, *Leçons sur les propriétés extrémales et la meilleure approximation des fonctions analytiques d'une variable réelle* (Gauthier-Villars, Paris, 1926), pp. 190 ff.
19. D. V. WIDDER, 'Functions whose even derivatives have a prescribed sign', *Proc. Nat. Acad. Sci. U.S.A.* 26 (1940) 657–659.
20. A. CLAUSING, 'Pólya operators. II. Complete concavity', *Math. Ann.* 267 (1984) 61–81.

Appendix 3

Comments on Number Theory

D. H. Lehmer

Pólya knew quite a lot of number theory and remained interested in the subject. As late as 1968, at the age of 80, he contributed an article about the sign of the error term in the prime number theorem to a collection of papers in honor of Edmund Landau [1968, 2]. However, only a few of his papers were entirely number theoretic. In reading his papers about power series with integer coefficients, for example, one is struck with the occasional applications of a number theory principle or a theorem from number theory.

Pólya was especially interested in the Legendre symbol (n/p) which is 1 or -1 according as n is a quadratic residue or a non-residue of an odd prime p. In 1912 Fekete announced the conjecture that the polynomial

$$\left(\frac{1}{p}\right)+\left(\frac{2}{p}\right)x+\left(\frac{3}{p}\right)x^2+\cdots+\left(\frac{p-1}{p}\right)x^{p-2}$$

has no real root between 0 and 1. Pólya [1919, 2] showed six years later that the conjecture was false for $p = 67$ and for infinitely many other primes. In the same paper he announced the conjecture bearing his name:

The excess of the number of integers $\leq x$ which have an odd number of prime factors over the number of integers with an even number of prime factors is non-negative.

In symbols: $L(x) \leq 0$ for $x > 1$, where $L(x) = \sum_{n \leq x} \lambda(n)$ and where $\lambda(n)$ is the Liouville function

$$\lambda\left(p_1^{\alpha_1}\cdots p_t^{\alpha_t}\right)=(-1)^{\alpha_1+\cdots+\alpha_t}.$$

Reprinted by permission from the *Bulletin of the London Mathematical Society* 19 (1987), 559–608.

Pólya himself verified his conjecture for $x \leq 1500$. This conjecture had a life of 40 years. By the 1950s it had been tested to 10^6 and for isolated values of x well beyond 10^6. Then in 1958 Haselgrove [2] disproved the conjecture by showing the existence of infinitely many x for which $L(x) > 0$. Not satisfied by this existence theorem, R. S. Lehman [3] found in 1960 that $L(906180359) = 1$. In 1980 Tanaka [4] examined all the numbers less than 10^9 and found the least failure of the Pólya conjecture to be 906150257.

Returning to the Legendre symbol, Pólya's name is attached to the useful inequality [1918, 3]

$$\left| \sum_{m=a}^{b} \left(\frac{m}{p} \right) \right| < \sqrt{p} \log p$$

In fact this is easily extended to

$$\sum_{m=a}^{b} \chi(m) = O\left(\sqrt{k} \log k \right)$$

where $\chi(m)$ is any non-principal Dirichlet character modulo k.

As is well known, Pólya was very much concerned with plausible and heuristic reasoning. When visiting Berkeley in 1959 he became aware of the results of an unpublished study that I had made of the number $\Pi_d(x)$ of pairs of primes $(p, p+d)$ for $x < 37 \cdot 10^6$. As a consequence he wrote an informal account [1959, 3] based on these data of the steps in a chain of reasoning that led him to the formula

$$\Pi_d(x) = 2C_2 \prod_{p|d} \left(\frac{p-1}{p-2} \right) \frac{x}{(\log x)^2},$$

where C_2 is the 'twin prime constant'

$$C_2 = \prod_{p>2} \left(1 - \frac{1}{(p-1)^2} \right) = 0.6601618....$$

This formula was conjectured 30 years earlier by Hardy and Littlewood [1]. The paper ends with this moral: 'Mathematicians and physicists think alike; they are led, and sometimes misled, by the same pattern of plausible reasoning.'

References

1. G. H. HARDY and J. E. LITTLEWOOD, 'Some problems of "Partitio Numerorum". III. On the expression of a number as a sum of primes,' *Acta Math.* 44 (1923) 1–70.

2. C. B. HASELGROVE, 'A disproof of a conjecture of Pólya,' *Mathematika* 5 (1958) 141–145.
3. R. SHERMAN LEHMAN, 'On Liouville's function', *Math. Comp.* 14 (1960) 311–320.
4. MINORU TANAKA, 'A numerical investigation on cumulative sum of the Liouville function,' *Tokyo J. Math.* 3 (1980) 187–189.

Appendix 4

Pólya's Geometry

Doris Schattschneider

One of Pólya's greatest strengths was his ability to recognize the importance of geometry in solving a variety of problems. His appreciation for
and use of geometry is found throughout his work.

One of his earliest papers [1913, 5] exploits the geometry of a scalene
right triangle to construct a Peano curve in which the unit interval [0, 1]
is continuously mapped onto the triangle and its interior. The construction is not well-known, yet is exceedingly easy to describe, and provides
a lovely example of a branching algorithm. A scalene right triangle has
the property that the altitude from its right angle to its hypotenuse splits
the triangle into two smaller right triangles T_1, T_2 of unequal size, both
of which are similar to the original triangle. The foot of that altitude is
the apex of altitudes that split each of T_1 and T_2 into two similar but
unequal triangles; and so on. Each $a \in [0, 1]$ can be represented in dyadic (binary) form as $0.a_1a_2a_3...$, where each a_i is 0 or 1; thus
$a = \sum_{i=1}^{\infty} a_i 2^{-i}$. This representation of a determines a path along the infinite binary tree of altitudes constructed in the triangle. Start at the foot
of the first altitude; if $a_1 = 0$, choose the altitude of the smaller of T_1 and
T_2. Travel to the foot of that altitude, where a_2 determines the next choice
($a_2 = 0 \Rightarrow$ smaller; $a_2 = 1 \Rightarrow$ larger), and so on. Figure 1 (taken from Pólya's
paper) shows the beginning of the path determined by $a = 0.1101...$.
For each a, the path determined by any dyadic representation of a con-
verges to a unique point in or on the original triangle, and the function
that maps a to that point is the desired Peano curve. Pólya shows that the
curve passes at most three times through any point of the triangle.

Reprinted by permission from the *Bulletin of the London Mathematical Society* 19 (1987),
559–608.

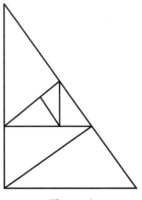

Figure 1

Geometric symmetry, and particularly the question of enumeration of symmetry classes of objects, was an early and continuing interest of Pólya. Although not the first to enumerate the 17 plane crystallographic groups (Fedorov had published this in 1891), Pólya was the first to illustrate each of these with a representative tiling [1924, 1]. Some tilings he took from classical sources and others he made up. In following his own familiar pedagogical maxim 'to make a picture,' he provided the reader with concrete figures in which to examine the differences in the symmetries of the tilings as well as to discern how, in a single tiling, the congruent copies of the tiles were related by symmetries. One reader of Pólya's paper for whom the illustrations conveyed all of the essential information was the Dutch artist M. C. Escher. The paper had been brought to Escher's attention by his brother, a geologist at the University of Leiden. In the three years 1937–39, Escher energetically produced over 25 colored periodic drawings of interlocked creatures, and corresponded with Pólya about his work [3].

Geometric symmetry, blended by Pólya with the theory of permutation groups and generating functions, produced one of his landmark theorems [1935, 4], and indeed, launched a whole theory which has come to be known as *Pólya enumeration*. The problem he attacked had come from the theory of *isomers* in organic chemistry, individual chemical species with identical molecular formulæ but displaying differing physiochemical properties (such as different arrangements of atoms). Chemists had sought, but failed to find, an algebraic technique that would enumerate isomerism classes of an organic chemical compound. A simple example (Figure 2) is the enumeration of the potential isomers formed when benzene (which has a hexagonal molecular 'frame') is substituted by a univalent radical *.

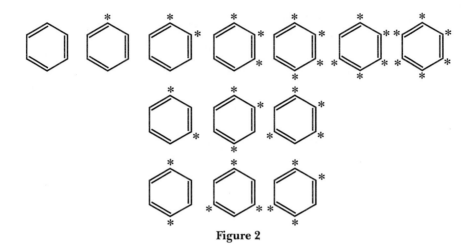

Figure 2

We can explain Pólya's theorem by obtaining this enumeration algebraically, using 'Pólya enumeration.' If G is a group of permutations on n symbols, the *cycle index* of G is a polynomial in n variables $x_1,\ldots,x_n$ which encodes information on how elements of G can be written as a product of disjoint cycles. Each element $\pi \in G$ determines a monomial $m(\pi)$ in the variables $x_1,\ldots,x_n$ as follows. Write π as a product of disjoint cycles, and to each cycle of length i associate the variable x_i; $m(\pi)$ is the product of these variables. The cycle index of G is the polynomial $P_G(x_1,\ldots,x_n) = |G|^{-1}\sum_{\pi \in G} m(\pi)$. For example, if G is the symmetry group of a regular hexagon (with vertices numbered cyclically $1, \ldots, 6$), then G is the dihedral group D_6 and the element $\pi = (135)(246)$, which corresponds to a rotation of $120°$, has $m(\pi) = x_3^2$. The cycle index for D_6 is

$$\tfrac{1}{12}\left(x_1^6 + 4x_2^3 + 3x_1^2x_2^2 + 2x_3^2 + 2x_6\right).$$

Labels are attached to the atoms of a molecular structure, and the figure-counting generating function records the number of ways in which this can be done. Pólya's theorem states that the isomerism enumeration generating function is obtained by substituting in the cycle index (of the permutation group that leaves invariant the molecular frame) the figure-counting generating function. We illustrate with the benzene example (Figure 2) in which there is one type of substituent [so the vertices of the hexagonal frame have no label (w), or have the * label (b)]. In this case each variable x_i in the cycle index for D is replaced by $b^t + w^t$ from the figure-counting series. The function so obtained is

$$\tfrac{1}{12}\left[(b+w)^6 +4(b^2 +w^2)^3 +3(b+w)^2(b^2 +w^2)^2 +2(b^3 +w^3)^2 +2(b^6 +w^6)\right]$$
$$= b^6 +b^5 w+3b^4 w^2 +3b^3 w^3 +3b^2 w^4 +bw^5 +w^6,$$

whose coefficients give the numbers of distinct isomers.

The far-reaching implications (and applications) of this theorem were apparent to Pólya, who gave a lengthy discussion and proof in the mathematical literature [1937, 3], and in addition gave explanations and illustrations of the enumeration method in several other scientific journals [1936, 1, 2, 3, 4]. Two recent papers by chemists describe the importance of Pólya's result and its many extensions (see [2] and [4]).

The role of geometric symmetry in minimizing geometric quantities and the resulting analogous minimization of physical quantities is lucidly explained by Pólya in his chapter 'Circle, sphere, symmetrization and some classical physical problems' [1961, 1] in a text for engineering students. He begins with the observation of Lord Rayleigh that of all membranes of equal area, the circle has not only the shortest perimeter, but also the lowest principal frequency. He then shows how the process of Steiner symmetrization of a geometric figure, while preserving area or volume, has a profound minimizing effect on physical quantities such as the Dirichlet integral, the principal frequency, and electrostatic capacity. The analogy between isoperimetric problems and certain physical quantities is also considered by Pólya in [1948, 4; 1951, 2; 1954, 4].

A very different sort of geometric problem, that of sightlines in an orchard, dates from an early paper [1918, 5] and appears as problem 239 in Pólya and Szegő's *Problems and theorems in analysis* [1925, 3]. A given circular orchard consists of trees of uniform thickness which are planted in a uniform array (like an integer lattice). How thick must the trees grow if, from the center of the orchard, every tree is visible, but the view beyond the orchard is completely blocked? A very nice account of the problem which gives a precise solution (Pólya gave only upper and lower bounds on the radius of the trees) can be found in [1]. Geometric symmetry plays a small role here as well—only a 45° wedge of the orchard need be considered.

Pólya used geometry in at least two distinct ways in his various pedagogical writings (notably, [1945, 2] and [1962, 1]). First, geometry was a favorite source of illustrative examples from which he could clearly demonstrate his teaching and problem-solving maxims. His generous use of figures shows the dynamic process of solution of geometric problems: look at a simple or special case, examine part of the problem or a related problem, 'turn it over and over, consider it under various aspects, study all sides' [1962, 1; Vol. 1, p. 111], look at it in a higher (or lower) dimension, embed it in a familiar figure, transform it and obtain information from the transformation process or from the transformed state. A second, quite different, use of geometry by Pólya was as metaphor and in representation of the solution process itself. Chapter 7 of *Mathematical Discovery* is entitled 'Geometric Representation of the Progress of the Solution.' Here the geometric words *connection, bridge, chain* and *thread* serve as descriptions of the ways in which solutions are built. What follows then is a detailed solution of a problem (in solid geometry) from which Pólya develops a schematic diagram (a digraph!) representing the multilevel process and progress of solution. This schematic representation of the problem and its solution is emblazoned on the end papers of the book.

References

1. T. T. ALLEN, 'Pólya's orchard problem,' *Amer. Math. Monthly* 93 (1986) 98–104.
2. A. T. BALABAN, 'Symmetry in chemical structures and reactions,' *Comput. Math. Appl.* Part B 12 (1986) 999–1020.
3. D. SCHATTSCHNEIDER, 'M. C. Escher's classification system for his colored periodic drawings,' *M. C. Escher: Art and Science* (H. S. M. Coxeter, M. Emmer, R. Penrose and M. L. Teuber, eds., North-Holland, 1986), pp. 82–96.
4. Z. SLANINA, 'An interplay between the phenomenon of chemical isomerism and symmetry requirements: a perennial source of stimuli for molecular-structure concepts, as well as for algebraic and computational chemistry,' *Comput. Math. Appl.* Part B 12 (1986) 585–616.

Appendix 5

Pólya's Enumeration Theorem

R. C. Read

During his long life George Pólya made notable contributions to many different branches of mathematics; but to combinatorialists he is chiefly known for his enumeration theorem—the 'Hauptsatz' of his 1937 paper [1937, 3] or [1987, 1]. This paper is remarkable in many ways, not the least being that it is a long paper devoted almost exclusively to a single theorem and its applications.

Pólya's theorem, as it is generally called, solves a very general type of combinatorial problem, which can be expressed in everyday terms as follows. Suppose we have a number of 'boxes' and a store of objects—called 'figures'—exactly one of which is to be placed in each box. (It may happen that two or more boxes receive the same figure.) The resulting structure—boxes plus figures—is called a 'configuration.' We further suppose that every figure has a 'content,' usually a non-negative integer, and we define the content of a configuration to be the sum of the contents of the figures in the boxes. Now, if the boxes are all distinct, it is a simple matter to determine the number of configurations having a given content.

What will happen if the boxes are not all distinct? In that case a certain rearrangement of the boxes in a configuration will give a new configuration which is equivalent, in some sense, to the original. More precisely, we have a certain group G of permutations of the boxes, and we say that two configurations are equivalent if one can be obtained from the other by permuting the boxes by some element of G. We then ask 'What is the number of inequivalent configurations having given content?' This is the problem to which Pólya's theorem provides the answer.

Reprinted by permission from the *Bulletin of the London Mathematical Society* 19 (1987), 559–608.

Pólya made elegant use of counting series and generating functions in his paper. The 'figure counting series' for a Pólya type problem is the power series in which the coefficient of x^n is the number of figures having content n; this is usually known from the statement of the problem. The 'configuration counting series' is defined similarly for configurations of content n; this series therefore summarizes all the answers to the problem. What is the connection between these two series that enables us to calculate the second from the first?

Clearly it must depend on the group G. Any permutation can be expressed as a product of disjoint cycles, and this expression is unique apart from the order of the cycles. For every permutation g of G we form a monomial $s_1^{j_1} s_2^{j_2} \cdots$ where j_i is the number of cycles of length i in g. The average of these monomials over all elements of G is what Pólya called the 'cycle index' of G. Pólya's theorem then states that the configuration counting series is obtained by substituting the figure counting series in the cycle index, by which is meant that if we denote the figure counting series by $f(x)$, we replace every occurrence of s_i in the cycle index by $f(x^i)$.

This is the theorem that makes possible the routine solution of a wide range of problems of both theoretical interest and practical importance. Let us look at some of these applications.

A typical application is seen in the enumeration of rooted trees. In a rooted tree, one vertex—the root—is distinguished from the others; suppose there are k edges incident with this vertex. At the other end of these edges we have again a rooted tree. Thus we have k 'boxes', in each of which we can place a rooted tree. In the basic problem of this kind the edges at the root, and hence the boxes, can be permuted in any way, and hence, in applying Pólya's theorem we take the group to be the full symmetric group S_k. The theorem then gives us the counting series for these trees (with k edges at the root) in terms of the counting series, $T(x)$, say, for all rooted trees. Summing this result over all values of k, and allowing for the extra vertex, the root, we reconstruct the counting series $T(x)$, which is then defined recursively. Such recursive definitions of a counting series are common in many applications of Pólya's theorem, and though they rarely give rise to an explicit formula for the number of configurations being enumerated, they are usually quite suitable for the numerical calculations of these numbers.

Chemical compounds can be represented by their structural formulæ which, to the graph theorist, are simply rather special types of graphs. In particular, acyclic chemical compounds, having no rings, correspond to trees. By methods similar to that outlined in the last paragraph, Pólya carried out the enumeration of various kinds of acyclic compounds, such as the alkanes (or paraffins), substituted alkanes, and many other families of compounds, determining the numbers of isomers (different compounds having the same numbers of atoms of various kinds). By appropriately choosing the group G he was able to effect this enumeration for the case where the shape of the molecule was taken into account (stereoisomers) as well as when it was not. In his 1937 papers and in some other papers published in preceding years [1935, 4; 1936, 2; 1936, 3; 1936, 4] Pólya also enumerated some kinds of cyclic chemical compounds, obtained by adding alkyl radicals, or other tree-like figures, to the atoms of various simple ring structures.

These and many other applications of the theorem make up the bulk of Pólya's paper. In the last section, however, Pólya applied his considerable analytical powers to derive asymptotic results for the many enumerative problems solved in the earlier sections. In so doing he was paving the way for much of the asymptotic enumeration that was to be carried out by later research workers.

Pólya applied his theorem to many problems besides those set out in his main paper. In 1940 he used it to solve a problem in logic [1940, 1], and it is known that he had successfully enumerated unlabelled graphs with given numbers of vertices and edges. Strangely enough, he never published his work on this problem, although it is an elegant and practical application of this theorem. This enumeration can be effected as follows. For graphs with a given number, p, of vertices, we regard every pair of vertices in the graph as being a 'box' into which we can put one of two figures, namely 'edge' or 'no edge,' with contents 1 or 0 respectively. Since the graphs are unlabelled there are no distinctions between vertices, which can therefore be permuted by any element of the symmetric group S_p. These permutations induce a group $S_p^{(2)}$ of the pairs of vertices, that is, of the boxes for this problem. The cyclic index of $S_p^{(2)}$ can be computed without too much trouble, and the required enumeration then follows directly by means of Pólya's theorem. Although Pólya did not publish this result himself, he communicated it to Harary, who published it in [2].

In the fifty years since the publication of Pólya's paper many advances have been made in the kind of enumerative combinatorics with which the paper was concerned. Chemists have found that Pólya's theorem can be used in their field, not just for finding the numbers of isomers of families of compounds, but for many other problems of a practical nature. Graph theorists with an interest in enumeration have made great use of Pólya's theorem, and in their endeavours to enumerate more and more complicated graphs have considerably extended Pólya's work. Generalizations of Pólya's theorem have been derived by de Bruijn [1], Harary and Palmer [3], Robinson [5] and many others. Indeed, in the last two or three decades enumerative graph theory has become a recognized branch of combinatorics (see [4] for further details).

This then was Pólya's outstanding contribution to the theory of enumeration, a remarkable theorem in a remarkable paper, and a landmark in the history of combinatorial analysis.

References

1. N. G. DE BRUIJN, 'Generalization of Pólya's fundamental theorem in enumerative combinatorial analysis', *Koninkl. Nederl. Akad. Wetensch.* A62 = *Indag. Math.* 21 (1969) 59–69.
2. F. HARARY, 'The number of linear, directed, rooted and connected graphs', *Trans. Amer. Math. Soc.* 78 (1955) 445–463.
3. F. HARARY and E. M. PALMER, 'The power group enumeration theorem', *J. Comb. Theory* 1 (1966) 157–173.
4. F. HARARY and E. M. PALMER, *Graphical enumeration* (Academic Press, 1973).
5. R. W. ROBINSON, 'Enumeration of acyclic digraphs', *Combinatorial mathematics and its applications* (R. C. Bose et al., eds., University of North Carolina, Chapel Hill, 1970), pp. 391–399.

Appendix 6

Pólya's Contributions in Mathematical Physics

M. M. Schiffer

Pólya's interest in complex analysis, conformal mapping and potential theory led him to the study of boundary value problems for partial differential equations and the theory of various functionals connected with them. This interest was strengthened by his teaching at the ETH in Zürich and at Stanford University which brought him in contact with engineers and their problems in classical physics and applied mathematics. In most cases boundary value problems for partial differential equations can be solved only in very special cases and for all other cases only approximate results can be obtained.

So, Pólya developed various techniques for estimating difficult functionals in terms of easier accessible quantities, be it other functionals or geometric quantities like area or volume. He obtained a large number of important and elegant inequalities and methods of approximation. Instead of enumerating such results, we shall concentrate on a few characteristic ideas to illustrate his approach.

Symmetrization. In many cases, boundary value problems for a given domain can be solved easily when the domain has a high symmetry; for example, when it is a circle or a sphere. From such a solution, general insights into the solutions for arbitrary domains can be obtained. As a typical example, let us take the case of the eigenvalues of a vibrating membrane. That is, consider a plane domain D with boundary C and the

Reprinted by permission from the *Bulletin of the London Mathematical Society* 19 (1987), 559–608.

partial differential equation $\nabla^2 u + \lambda u = 0$ with the boundary condition $u = 0$ on C. In general, the only solution is the trivial one: $u \equiv 0$. However, for a series of positive numbers $\lambda_1 \leq \lambda_2 \leq \lambda_3 \leq \cdots$, there exist solutions $u_\nu(x, y)$ which are not identically zero. These λ_ν are called the eigenvalues and the functions u_ν the eigenfunctions of the problem. These solutions are known only for a small number of domains D, in particular for the circle. In 1894 Lord Rayleigh conjectured that among all domains of given area the circle has the lowest eigenvalue λ_1 [4]. He justified this conjecture by considering all domains for which λ_1 was known and also by a simple variational argument. However, only in 1923 did Faber give the first proof of this conjecture [1], while in 1924 Krahn gave an independent proof [2]. On the basis of this result, the difficult eigenvalue λ_1 can be estimated from below in terms of the much more accessible geometric measure of area. This fact plays a role in acoustics and potential theory.

Inequalities between different domain functionals occurred already in classical antiquity. The most famous one stated that among all plane domains D with given area A, the circle has the least perimeter L; that is, for all domains D, we have the inequality $L^2 \geq 4\pi A$. This is the famous isoperimetric inequality, which may be considered as the first result in the calculus of variations. Because of it, we call now, in general, inequalities between various domain functionals 'isoperimetric inequalities.' Much effort was made to give a rigorous proof for the classical isoperimetric inequality. One of the most ingenious methods of proof is due to the Swiss mathematician Jakob Steiner (see [1961, 1] and [5]). He introduced the concept of symmetrization of a plane domain D with respect to a given line. Suppose that D lies in the (x, y)-plane over an interval $a \leq x \leq b$ of the abscissa. Through every point x in this interval, we draw a vertical line which intersects the domain D in one or more intervals of total length $\lambda(x)$. On another copy of the interval (a, b), we draw through each point x a vertical line of length $\lambda(x)$ which is centered on that axis, that is, $-\frac{1}{2}\lambda(x) \leq y \leq \frac{1}{2}\lambda(x)$. The endpoints of these segments determine a closed curve C^* which is symmetric with respect to the x-axis and which determines a domain D^* with the same symmetry. We call D^* the symmetrization of D with respect to the x-axis. It is easy to see that D^* has the same area as the original domain D but that the length L^* of its boundary is less than or equal to the perimeter L of D. Thus, the domain of a given area with the least perimeter must have the highest symmetry, that is, it must be a circle. Precisely the same method of symmetrization works in space with respect to a plane. It follows that among all bodies with given volume V, the sphere has the least surface area S, that is, we have the isoperimetric inequality: $V^2 \leq (1/36\pi)S^3$.

Now, Pólya made the ingenious remark that Steiner's method allows a large number of applications to the theory of partial differential equations (see [1948, 4] and [1961, 1]). Consider a domain D in the (x, y)-plane and a function $f(x, y)$ defined in D which vanishes on its boundary C. This determines a body B in the (x, y, z)-space which is bounded by the surface $z = f(x, y)$ and the plane piece D. Let Π be a plane which is perpendicular to the plane $z = 0$ and symmetrize B with respect to Π. This gives a body B^* of the same type which is bounded by a surface $z = f^*(x, y)$ and by a flat base D^* in the (x, y)-plane. Then B^* has the same volume as B but a lesser surface area. Observe also that D^* has the same area as D. By the calculus formula for the surface area, we have therefore:

$$\iint_D \left(1 + f_x^2 + f_y^2\right)^{1/2} dx\, dy \geq \iint_{D*} \left(1 + f_x^{*2} + f_y^{*2}\right)^{1/2} dx\, dy.$$

Now consider the function $\varepsilon f(x, y)$ with an arbitrary smallness parameter $\varepsilon > 0$. The above inequality holds again and, by passage to the limit $\varepsilon \to 0$, we conclude that

$$\iint_D \left(f_x^2 + f_y^2\right) dx\, dy \geq \iint_{D*} \left(f_x^{*2} + f_y^{*2}\right) dx\, dy.$$

Now, the integral

$$D[f] = \iint_D \left(f_x^2 + f_y^2\right) dx\, dy$$

occurs in many applications in classical physics as the Dirichlet energy integral. Thus, Pólya has found a method of symmetrization for the Dirichlet integral in a domain D for arbitrary functions $f(x, y)$ in it which vanish on its boundary C.

An immediate application of this idea was a new elegant proof of Rayleigh's conjecture. Another was a proof of a conjecture of Poincaré that among all conducting surfaces that enclose a given volume, the sphere has the least electrostatic capacity [3]. This had already been proved by G. Szegő by an ingenious argument [6], but now it followed quite easily from a general method.

A large number of further inequalities relating to Dirichlet's integral could be obtained. In collaboration with Szegő, Pólya wrote a monograph *Isoperimetric inequalities in mathematical physics* [1951, 2] which is now a classic and has stimulated many further investigations in this field. Combining various methods with great analytical skill the authors obtain estimates and inequalities for numerous functionals. We mention capacity, torsional rigidity, virtual mass, polarization, transfinite diameter and interior radius. Various new kinds of symmetrizations are invented; for example, circular symmetrization with respect to a given ray. The whole work displays the taste of the authors for the concrete and explicit result, for elegance and ingenious methods.

Transplantation. Many functionals that occur in classical physics and engineering can be characterized by extremum problems in appropriate function spaces. Consider, for example, the torsional rigidity P of a plane domain D. It is defined in terms of the stress function $f(x, y)$ of D which satisfies in D the partial differential equation $\nabla^2 f + 2 = 0$ and which vanishes on the boundary C of D. In terms of $f(x, y)$, the functional P is given as

$$P = \left(2 \iint_D f(x, y) dx dy\right)^2 \cdot \left(\iint_D (\nabla f)^2 dx dy\right)^{-1}.$$

It is then easily seen that for every continuously differentiable function $F(x, y)$ in D which vanishes on C, we have

$$P \ge \left(2 \iint_D F(x, y) dx dy\right)^2 \cdot \left(\iint_D (\nabla F)^2 dx dy\right)^{-1}.$$

Thus, we can define P directly by means of a maximum problem within a given function class and can use every element in that class to obtain a lower bound for it.

But we can use this characterization to a much greater advantage. We may imbed the given domain D in a one-parameter family of domains $D(t)$, where $D(t)$ is obtained from $D = D(1)$ by stretching of the entire (x, y)-plane $x' = tx$, $y' = y$, $t > 0$. Now let $f(x, y)$ be the correct stress function of the domain $D(t_0)$. Then $f(t_0 t^{-1} x, y)$ is defined in $D(t)$, vanishes on the boundary $C(t)$ and is thus an admissible competing function in the extremum problem which defines $P(t)$. We call this function the transplant from $D(t_0)$ to $D(t)$ of the extremum function for $D(t_0)$. An easy calculation shows that the maximum property of $P(t)$ implies

$$\frac{t}{P(t)} \le t_0 \frac{\tau t_0^2 \iint_{D(t_0)} f_x(x, y)^2 dx dy + \iint_{D(t_0)} f_y(x, y)^2 dx dy}{\left(2 \iint_{D(t_0)} f(x, y) dx dy\right)^2}$$

with $\tau = t^{-2}$. The right-hand-side of this estimate has the form $a\tau + b$ and for $\tau_0 = t^{-2}$ we have equality in this estimate. Let us plot the term $tP(t)^{-1}$ versus the abscissa τ; we can assert that through each point on that curve passes a straight line such that all curve points lie below it. Thus this curve is convex and these straight lines are its supporting lines. The slope of the supporting line at the point t_0 is

$$a = \tau_0^{-3/2} \iint_{D(t_0)} f_x^2 \, dx \, dy \cdot P(t_0)^{-2} \geq 0,$$

so that the curve is non-decreasing. If in the family there is a domain $D(t_0)$ for which the stress function is known, we obtain estimates for all $P(t)$.

This example shows clearly the general idea of how we can utilize the extremum characterization of a functional to study the parameter dependence of the functionals for a parameter-dependent family of domains. One uses the extremum function of a given domain in the family as a test function for all other domains in the family by proper transplantation. This idea was carried out in many cases in an extensive paper in collaboration with M. Schiffer [1954, 1]. The most useful extremum problems considered were the Dirichlet and the Thomson principles. Since these principles give, in general, lower and upper bounds, respectively, for the functionals considered, they work very well in combination to bound the functional from below and above. The functionals discussed in the paper were, for example, virtual mass, capacity, membrane eigenvalues and outer radius. The paper stimulated much further research along these lines and was followed by a number of extensions and generalizations of its results and methods. In particular, J. Hersch and his group at the ETH in Zürich were especially active and successful in this direction.

The Method of Difference Equations. Many boundary value problems in the theory of partial differential equations can be solved numerically and approximately by studying appropriate difference equations. Pólya made important contributions in this connection. He combined the approximation by difference equations in an ingenious way with the Rayleigh-Ritz method (see [1952, 3] and [1954, 3]). He interpolated the discrete function values obtained by the difference equation method at the grid points in a linear or bilinear way. Thus he obtained functions defined at all points in the domain considered and used these functions as particularly good test functions in the Rayleigh-quotient. These ideas are closely related to the method of finite elements which has developed to a favorite tool in applied and engineering mathematics. Again he combined various general techniques to obtain upper and lower bounds for the eigenvalues in question.

Space does not permit enumerating his many further ideas, conjectures and methods in this field. But we should mention the number of his colleagues and collaborators whom he has attracted to this subject and who under his inspiration have continued in his tradition: to mention only a few, Hersch, Pfluger, Payne, Weinberger and Weinstein.

References

1. G. FABER, 'Beweis, dass unter allen homogenen Membranen von gleicher Fläche und gleicher Spanning die kreisförmige den tiefsten Grundton gibt', *Sitz. Bayr. Akad.* (1923) 169–172.
2. E. KRAHN, 'Über eine von Rayleigh formulierte Minimaleigenschaft des Kreises', *Math. Ann.* 94 (1924) 97–100.
3. H. POINCARÉ, *Figures d'équilibre d'une masse fluide* (Paris, 1903).
4. J. W. G. RAYLEIGH, *The theory of sound*, Vol. 1 (London, 1894).
5. J. STEINER, *Gesammelte Werke* (Berlin, 1881–1882).
6. G. SZEGŐ, 'Über einige Extremalaufgaben der Potentialtheorie', *Math. Z.* 31 (1930) 583–593.

Appendix 7

George Pólya and Mathematics Education

Alan H. Schoenfeld

It has been noted that ordinary mortals can see a long way when they stand on the shoulders of giants. Pólya invoked this image in acknowledging his debt to Descartes, whose introspections about his own mathematical thinking (see, for example, Descartes' *Rules for the Direction of the Mind*) served as a major inspiration for Pólya's thoughts on the topic (see, for example, *How to solve it,* the two volumes of *Mathematics and plausible reasoning,* and the two volumes of *Mathematical discovery*). This homage was made with Pólya's typical humility. The fact is that those who explore the nature of mathematical thinking stand atop a pyramid of giants—and their feet are firmly set on Pólya's shoulders.

How strong is Pólya's influence? Education lies in the public arena, so there are two aspects of that influence to explore: (1) the impact of Pólya's work and ideas in the real world, and (2) the solidity of his work as a base for making scientific progress on issues related to understanding and teaching the nature of mathematical thinking. We consider both in order.

How to solve it, Pólya's first book-length foray on heuristics and education, appeared in 1945. (This was hardly his first educational work: Pólya and Szegő's *Aufgaben und Lehrsätze aus der Analysis I* appeared in 1925, and it was preceded by writings on heuristics.) The flyleaf of *How to solve it* contained the outline of Pólya's four-stage approach to the problem solving process: understanding the problem, devising a plan, carrying

Reprinted by permission from the *Bulletin of the London Mathematical Society* 19 (1987), 559–608.

out the plan, and looking back (checking the solution). More than three decades later, the National Council of Teachers of Mathematics (NCTM) declared, in its 1980 *Agenda for Action* [3], that 'Problem solving must be the focus of school mathematics in the 1980s'. To help this process along, NCTM devoted its 1980 Yearbook [2] to *Problem Solving in School Mathematics*. If you open the Yearbook you will find Pólya's four-stage approach to problem solving reproduced in its inside covers. Continue reading and you will find that the vast majority of articles are based on Pólya's ideas about mathematical thinking. Such obvious homage, combined with Pólya's position as honorary president of the Fourth International Congress on Mathematical Education (Berkeley, 1980), testify to Pólya's preeminence as a mathematics educator.

Moreover, Pólya's influence extends far beyond the mathematics education community. Just as the 'back to basics' movement in mathematics during the 1970s was symptomatic of the 'basic skills' movement cutting through education at large, we find that the 'problem solving movement' cuts a wide swath in the 1980s. In addition to the predictable citations of Pólya's work in the *American Mathematical Monthly,* the *Journal for Research in Mathematics Education*, and other journals with an emphasis on mathematics education, one also finds recent citations of Pólya's writing in the *American Political Science Review, Annual Review of Psychology, Artificial Intelligence, Computers and Chemistry, Computers and Education, Discourse Processes, Educational Leadership, Higher Education, Human Learning*—to name just a few.

The scientific status of Pólya's work on problem-solving strategies has been more problematic. While in general the quality of one's contributions to mathematics is pretty clear, the quality of one's contributions to the psychology of thinking is less so. (Consider, for example, the rises and falls of Freud's reputation through the years.) It is true that Pólya's writings on 'modern heuristic' have generally struck a resonant chord with mathematicians, and have inspired numerous mathematics educators to teach problem solving *via* heuristics—but it is also true that such attempts, for the most part, have had minimal success. The math-ed literature is chock full of heuristic studies with 'promising' results. That is, students and instructors alike felt that a heuristics-based approach to course work was worthwhile, but there was rarely convincing evidence to show that the students' problem-solving performance had actually improved as a result of that approach. On the basis of instructional results, Pólya's theoretical ideas can be challenged.

Perhaps more importantly, those ideas have been challenged by a set of competing ideas from another discipline. In contrast to the fuzziness of qualitative psychology and of some educational experimentation, researchers in artificial intelligence (AI) offered what they would call real science. If one adopts the hardnosed AI point of view, no statement about cognition is proved until you have a runnable computer program that embodies that statement—so no problem solving theory is accepted until you have a program that solves problems using the theory. By that standard, Pólya's ideas fall short. As one leading AI researcher put it, 'We tried to write problem solving programs using Pólya's heuristics, and they failed; we tried other methods, and they succeeded. Thus we suspect the strategies he describes are epiphenomenal rather than real—and even if they are real, they're far less important than the ones we use in our programs.'

In both mathematics education and in AI, then, there has been empirical reason to question the solidity of the foundations established by Pólya. In recent years, however, there is increasing evidence of the solidity of those foundations. There is reason to believe that on both counts—in mathematics education and in artificial intelligence—the next decade will swing scientific opinion back in Pólya's direction. In essence, the difficulties with the implementation of Pólya's ideas were that (a) they were not specified in adequate detail for implementation, and (b) they appeared to be superseded by more 'general' methods. Recent work in cognitive science has provided the means of addressing both of these issues. First, cognitive science has provided methods for fleshing out the details of Pólya's strategies, making them more accessible for problem solving instruction. There are now studies providing clear evidence that students can learn problem solving *via* heuristics, with significant improvements in their problem-solving performance. (See, for example, [4].) In addition, the general methods of AI have turned out to be much weaker than had been thought; methods once thought general and powerful have turned out to have limited scope and power. Research from the past decade indicates that problem-solving strategies are much more tightly bound to domain-specific subject matter understandings than early AI researchers had claimed. In consequence, current research focuses on the elaboration of problem-solving strategies tied to bodies of subject matter. With increased sophistication in characterizing 'the knowledge structures required to operate on semantically rich domains' (for example, mathematical problem solving), the field has reached the point where it may be possible to program computer-based knowledge structures capable of supporting heuristic problem-solving strategies of the type Pólya described. Should that be the case—and this author predicts it will—research will provide the tools to implement Pólya's intuitions about problem solving, which will serve as part of the foundation for a true 'science of thought'.

References

1. R. DESCARTES, *Rules for the direction of the mind* (E. S. Haldane and G. R. I. Ross, translators), Great Books of the Western World, Vol. 31 (Chicago, Encyclopedia Britannica Inc., 1952).
2. S. KRULIK (Ed.), *Problem solving in school mathematics,* 1980 Yearbook of the National Council of Teachers of Mathematics (National Council of Teachers of Mathematics, 1980).
3. National Council of Teachers of Mathematics, *An agenda for action* (National Council of Teachers of Mathematics, 1980).
4. A. SCHOENFELD, *Mathematical problem solving* (Academic Press, 1985).

Postscript added July 16, 1999:

Over the years since this essay was written, mathematics education and artificial intelligence (AI) have evolved in interesting and somewhat independent ways. As predicted, AI has focused more on domain-specific knowledge and less on general strategies. In that sense, Pólya's ideas were right on the mark–although it should be noted that there is little AI work on general mathematical problem solving these days, and much more on specific engineering applications.

It is fair to say, I think, that Pólya's spirit lives on in mathematics education. Indeed, current practices may be more in line with Pólya's ideas than the early, somewhat simplistic attempts to teach problem solving strategies as isolated tools and techniques. The National Council of Teachers of Mathematics' 1989 *Curriculum and Evaluation Standards* for School Mathematics defined not only "content (i.e., subject matter) standards" for grades K–12, but highlighted these four "process standards" as well: problem solving, reasoning, communications, and connections. The shift, which has not been uncontroversial, has been toward mathematics instruction that focuses as much on these aspects of mathematical thinking as it does on core knowledge of mathematics. To the degree that one can keep a solid mathematical core and focus on having students truly engage with and do mathematics, this shift does reflect Pólya's deep ideas about mathematical thinking and problem solving.

Alan H. Schoenfeld

Appendix 8

Pólya's Influence—
References to His Work

Pólya added considerably to the mathematical language through phrases like "random walk" and "central limit theorem." But he has also indirectly added to our mathematical vocabulary by having named for him a number of concepts that he introduced. Of course, Boyer's Law tells us that if something is named for someone, that person is probably not the person who first observed it. This is clear in Pólya's case too: an example is the Pólya Enumeration Theorem which was first observed by J.H. Redfield. Nevertheless, rigorous historical scholarship aside, it is interesting to observe how many phrases have entered the literature where credit is given to Pólya.

The first version of this list appeared in *Mathematics Magazine* 60 (1987), 259–63; what follows below is a considerably updated version. The list is not exhaustive and references are not necessarily the definitive introduction to the concept but provide entree to the subject.

Gauss-Pólya Inequality
  (See Pólya Inequality.)

Hardy-Littlewood-Pólya Inequality
  Ke Hu, On Hardy-Littlewood-Pólya inequality and its applications. *Chinese J. Contemp. Math.* 15 (1994), 325–330.

Hardy-Littlewood-Pólya-Everitt Inequality
  Christer Bennewitz, A general version of the Hardy-Littlewood-Pólya-Everitt inequality, *Proc. Roy. Soc. Edinburgh* Sect. A 97 (1984), 9–20.

Laguerre-Pólya Functions
  (See Pólya-Schur Functions.)

Lindwart-Pólya Theorem
S. Hellerstein and J. Korevaar, Limits of entire functions whose growth and zeros are restricted, *Duke Math. J.* 30 (1963), 221–228.

Lupas-Pólya Operators
Wen Zhong Chen, The order of convergence of sequences of Lupas-Pólya operators, *Xiamen Daxue Xuebao Ziran Kexue Ban* 35 (1996), 309–314.

Markov-Pólya Distributions
K. G. Janardan and B. Raja Rao, Characterization of generalized Markov-Pólya and generalized Pólya-Eggenberger distributions, *Comm. Statist. A—Theory Methods* 11 (1982), 2113–2124.

Nevanlinna-Pólya Theorem
Hiroshi Haruki, A generalization of the Nevanlinna-Pólya theorem in analytic function theory, *Math. Notæ* 29 (1981), 29–35.

Payne-Pólya-Weinberger Conjecture
Mark S. Ashbaugh and Rafael D. Benguria, A second proof of the Payne-Pólya-Weinberger conjecture, *Comm. Math. Phys.* 147 (1992), 181–190.

Payne-Pólya-Weinberger Inequality
G. Pólya, *Collected Papers*, Vol. 3, MIT Press, 1984, p. 519.

Peano-Pólya Curve
(See Pólya Curve.)

Peano-Pólya Motions
B. B. Mandelbrot and Stéphane Jaffard, Peano-Pólya motions, when time is intrinsic or binomial (uniform or multifractal), *Math. Intelligencer* 19 (1997), 21–26.

Plancherel-Pólya-Nikol'skij Inequality
Bernd Stöckert, Ungleichungen vom Plancherel-Pólya-Nikol'skij-Typ in gewichteten L_p^{Ω}-Räumen mit gemischten Normen, *Math. Nachr.* 86 (1978), 19–32.

Pólya Algorithm
Andras Kroó, On the convergence of Pólya's algorithm, *J. Approx. Theory* 30 (1980), 139–148.

Pólya-Bernstein Theorem
Maurice Blambert and Rajagopalan Parvatham, Compléments à des

théorèmes de S. Mandelbrot et de Pólya-Bernstein, *C. R. Acad. Sci. Paris* Ser. A-B 290 (1980), A457–A460.

Pólya-Cantor Theorem
  Khira Lameche, Extension d'un théorème de Pólya-Cantor à des séries rationnelles en variables non commutatives, *Séminaire Delange-Pisot-Pointou,* 1970–71.

Pólya-Carlson Theorem
  Ralph Boas, *Entire Functions,* Academic Press, 1954, p. 178.

Pólya Characteristic Functions
  A. I. Il'inskii, The arithmetic of Pólya characteristic functions, *Mat. Zametki* 21 (1977), 717–725.

Pólya Conditions
  G. Pólya, *Collected Papers,* Vol. 3, MIT Press, 1984, p. 494.

Pólya Conjecture
  Harold M. Stark, *An Introduction to Number Theory,* Markham, 1970, p. 7.

Pólya Counting Theorem
  (See Pólya Enumeration Theorem.)

Pólya Criterion
  G. Pólya, *Collected Papers,* Vol. 4, MIT Press, 1984, p. 613.

Pólya Curve
  Phillip J. Barry and Ronald N. Goldman, Shape parameter deletion for Pólya curves. *Numer. Algorithms* 1 (1991), 121–137.

Pólya-de Bruijn Theorem
  Shaoqin Han, An exposition of three limitation problems connected with the Pólya-de Bruijn theorem, *J. Math. Res. Exposition* 9 (1989), 565–566.

Pólya Density
  Robert M. Berk, Some monotonicity properties of symmetric Pólya densities and their exponential families, *Z. Wahrsch. Verw. Gebiete* 42 (1978), 303–307.

Pólya Distribution
  William Feller, *An Introduction to Probability Theory and Its Applications,* 3rd ed., Vol. 1, Wiley, 1950, p. 142.

Pólya-Eggenberger Distributions
(See Markov-Pólya Distributions.)

Pólya Enumeration Theorem
Alan Tucker, Pólya's enumeration formula by example, *Math. Mag.* 47 (1974), 248–256.

Pólya Factorizations
Uri Elias, Integral means and Pólya factorizations, *Proc. Amer. Math. Soc.* 126 (1998), 2071–2075.

Pólya Frequency Functions
I. J. Schoenberg, On Pólya frequency functions I. The totally positive functions and their Laplace transforms, *J. Analyse Math.* 1 (1951), 331–374.

Pólya Function
(See Pólya Curve.)

Pólya Gap Theorem
T. Kovari, On the gap theorem of Pólya, *J. London Math. Soc.* 34 (1959), 185–194.

Pólya Inequality
C. E. M. Pearce, et al., A generalization of Pólya's inequality to Stolarsky and Gini means, *Math. Inequal. Appl.* 1 (1998), 211–222.

Pólya Logic
Zhen Hua Ma, Pólya logic. I, *J. Tsinghua Univ.* 28 (1988), 1–8.

Pólya-Lundberg Process
Dietmar Pfeifer, A note on the occurrence times of a Pólya-Lundberg process, *Adv. in Appl. Probab.* 15 (1983), 886.

Pólya-Macintyre Representation Theory
Paul Malliarin and L. A. Rubel, On small entire functions of exponential type with given zeros, *Bull. Soc. Math. France* 89 (1961), 175–206.

Pólya-Mammana Factorization
A. A. Aizikovich, The Pólya-Mammana factorization of a linear matrix difference operator. I, *Differencial-nye Uravnenija* 14 (1978), 328–337, 388–389.

Pólya Matrices
G. M. Peterson and Anne C. Baker, On a theorem of Pólya (II), *J. London Math. Soc.* 39 (1964), 745–752.

Pólya Means
  L. A. Rubel, Maximal means and Tauberian theorems, *Pacific J. Math.*
  10 (1960), 997–1007.

Pólya Operators
  Achim Clausing, Pólya operators. I. Total positivity, *Math. Ann.* 267
  (1984), 37–59.

Pólya Orchard Problem
  Thomas Tracy Allen, Pólya's orchard problem, *Amer. Math. Monthly* 93
  (1986), 98–104.

Pólya-Padé Fourier Resonance
  Carlos R. Handy, Pólya-Padé resonance reconstruction and singular
  perturbation theory, *Nonlinear Analysis* 10 (1986), 391–401.

Pólya Peaks
  W. K. Hayman, *Meromorphic Functions,* Oxford, 1964, p. 101.

Pólya Point Process
  Ed. Waymire and Vijay K. Gupta, An analysis of the Pólya point process, *Adv. in Appl. Probab.* 15 (1983), 39–53.

Pólya Polynomials
  Rudolf Land, Computation of Pólya polynomials of primitive permutation groups, *Math. Comp.* 36 (1981), 267–278.

Pólya Process
  Ci Wen Xu, Local nondeterminism and the uniform dimension of the
  inverse image of a Pólya process, *J. Wuhan Univ. Natur. Sci. Ed.* 41
  (1995), 526–532.

Pólya Property
  Chiu-Cheng Chang, On a class of Pólya property preserving operators, *Soochow J. Math.* 22 (1996), 495–511.

Pólya Representations
  Ralph Boas, *Entire Functions,* Academic Press, 1954, p. 74.

Pólya-Saint Venant Theorem
  G. Pólya, *Collected Papers,* Vol. 3, MIT Press, 1984, pp. 500–502.

Pólya-Schiffer Inequality
  G. Pólya, *Collected Papers,* Vol. 3, MIT Press, 1984, p. 519.

Pólya-Schoenberg Conjecture
St. Ruscheweyh and T. Sheil-Small, Hadamard products of Schlicht functions and the Pólya-Schoenberg conjecture, *Comment. Math. Helv.* 48 (1973), 119–135.

Pólya-Schur Functions
I. J. Schoenberg, On totally positive functions, Laplace integrals and entire functions of the Laguerre-Pólya-Schur type, *Proc. Nat. Acad. Sci. U.S.A.* 33 (1947), 11–17.

Pólya Sequences
David W. Walkup, Pólya sequences, binomial convolution and the union of random sets, *J. Appl. Probab.* 13 (1976), 76–85.

Pólya Series
Christophe Reutenauer, On Pólya's series in noncommuting variables, *Fundamentals of Computation Theory,* Berlin/Wendisch-Rietz, 1979.

Pólya-Sonine Theorem
J. Steinig, The real zeros of Struve's function, *SIAM J. Math. Anal.* 1 (1970), 365–375.

Pólya States
Hong-Chen Fu, Pólya states of quantized radiation fields, their algebraic characterization and non-classical properties, *J. Phys. A* 30 (1997), L83–L89.

Pólya-Szegő Composition Formula
Jacob Burbea, Total positivity of certain reproducing kernels, *Pacific J. Math.* 67 (1976), 101–130.

Pólya-Szegő Inequality
Motosaburo Masuyama, A refinement of the Pólya-Szegő inequality and a refined upper bound of CV, *TRU Math.* 21 (1985), 201–205.

Pólya-Szegő Matrices
A. B. Movchan and S. K. Serkov, The Pólya-Szegő matrices in asymptotic models of dilute composites, *European J. Appl. Math.* 8 (1997), 595–621.

Pólya Theory of Hypercubes
P. W. H. Lemmens, Pólya theory of hypercubes, *Geom. Dedicata* 64 (1997), 145–155.

Pólya Trees
Pietro Muliere, A Bayesian non-parametric approach to survival analysis using Pólya trees. *Scand. J. Statist.* 24 (1997), 331–340.

Pólya-Turán Inequalities
C. L. Prather, Some generalized Pólya-Turán inequalities, *J. Math. Anal. Appl.* 151 (1990), 140–163.

Pólya 2^z Theorem
G. Pólya, *Collected Papers*, Vol. 1, MIT Press, 1974, p. 771.

Pólya Urn Scheme
William Feller, *An Introduction to Probability Theory and Its Applications*, 3rd ed., Vol. 1, Wiley, 1950, p. 120.

Pólya-Vinogradov Inequality
Peter Söhne, The Pólya-Vinogradov inequality for totally real algebraic number fields, *Acta Arith.* 65 (1993), 197–212.

Pólya W-Property
Miseal Zedek, Cayley's decomposition and Pólya's W-property of ordinary differential equations, *Israel J. Math.* 3 (1965), 81–86.

Pólya-Weinstein Inequality
G. Pólya, *Collected Papers*, Vol. 3, MIT Press, 1984, p. 508.

Pólya-Wiman Conjecture
Young-One Kim, A proof of the Pólya-Wiman conjecture, *Proc. Amer. Math. Soc.* 109 (1990), 1045–1052.

Appendix 9

Prizes, Awards and Lectureships Honoring George Pólya

Pólya Prizes in Combinatorics
(Society for Industrial and Applied Mathematics)

These prizes, given every four years, were begun with the help of a gift from Frank Harary, to honor contributors to the field of combinatorics. The range of topics covered was expanded in 1992 to recognize "a notable contribution in another area of interest to George Pólya such as approximation theory, complex analysis, number theory, orthogonal polynomials, probability theory, or mathematical discovery and learning" and for combinatorial theory applications. The prizes are now awarded every two years. The medals are accompanied by a cash award of $20,000.

1971   Ronald L. Graham (AT&T Bell Labs), Klaus Leeb (University of Erlangen) , Bruce L. Rothschild (University of California, Los Angeles), Alfred W. Hales (University of California, Los Angeles), and Robert I. Jewett (Rio de Janeiro)

1975   Richard P. Stanley (Massachusetts Institute of Technology), Endre Szemerédi (Hungarian Academy of Sciences), and Richard M. Wilson (California Institute of Technology)

1979   László Lovász (Yale University and Eötvös University)

1983   Anders Björner (Massachusetts Institute of Technology, Swedish Royal Institute of Technology) and Paul Seymour (BELLCORE)

1987 Andrew Chi Chih Yao (Princeton University)
1992 Gil Kalai (Hebrew University and IBM, San Jose) and Saharon Shelah (Hebrew University)
1994 Gregory Chudnovsky (Columbia University) and Harry Kesten (Cornell University)
1996 Jeffry Ned Kahn (Rutgers University) and David Reimer (Middlesex Community College)
1998 Percy Deift (New York University-Courant), Xin Zhou (Duke University), and Peter Sarnak (Princeton University)

Pólya Prizes
(London Mathematical Society)

These prizes are awarded "to an individual in recognition of outstanding creativity in, of imaginative exposition of, and of distinguished contribution to, mathematics within the United Kingdom." The Prize is given in years when the Society is not awarding the De Morgan Medal.

1987 John Horton Conway (University of Cambridge)
1988 Charles T. C. Wall (Liverpool University)
1990 Graeme Bryce Segal (University of Oxford))
1991 Ian Grant Macdonald (Queen Mary & Westfield College, London)
1993 David Rees (University of Exeter)
1994 David Williams (University of Bath)
1996 D. E. Edmunds (University of Sussex)
1997 J. M. Hammersley (University of Oxford)
1999 Simon Donaldson (Imperial College, London)

George Pólya Awards
(Mathematical Association of America)

These awards are for expository writing in the *College Mathematics Journal.*

1977 Anneli Lax (Courant Institute of Mathematical Sciences)
       Julian Weisglass (University of California, Santa Barbara)
1978 Allen H. Holmes (St. Paul Academy and Summit School), Walter J. Sanders (Indiana State University), and John W. LeDuc (Eastern Illinois University)
1979 Richard L. Francis (Southeast Missouri State University)
       Richard Plagge (Highline Community College)

1980   Hugh F. Ouellette (Winona State University) and Gordon Bennett
(Western Montana College)
Robert Nelson (Menlo School)

1981   Gulbank D. Chakerian (University of California, Davis)
Robert G. Dean (Stephen F. Austin University), Ennis D. McCune
(Stephen F. Austin University), and William D. Clark (Stephen F.
Austin University)

1982   John A. Mitchem (San Jose State University)
Peter L. Renz (Mathematical Association of America)

1983   Douglas R. Hofstadter (Indiana University)
Paul R. Halmos (Indiana University)
Warren Page (New York City Technical College) and V. N. Murty
(Pennsylvania State University, Capitol Campus)

1984   Ruma Falk (Hebrew University) and Maya Bar-Hillel (Hebrew
University)
Richard J. Trudeau (Stonehill College)

1985   Anthony Barcellos (State of California)
Kay W. Dundas (Hutchinson Community College)

1986   Philip J. Davis (Brown University)

1987   Constance Reid (San Francisco, California)
Irl C. Bivens (Davidson College)

1988   Dennis M. Luciano (Western New England College) and Gordon
D. Prichett (Babson College)
V. Frederick Rickey (Bowling Green State University)

1989   Edward Rozema (University of Tennessee at Chattanooga)
Beverly L. Brechner (University of Florida) and John C. Mayer
(University of Alabama, Birmingham)

1990   Israel Kleiner (York University)
Richard D. Neidinger (Davidson College)

1991   William B. Gearhart (California State University, Fullerton) and
Harris S. Shultz (California State University, Fullerton)
Mark Schilling (California State University, Northridge)

1992   William Dunham (Hanover College)
Howard Eves (University of Maine)

1993   Lester H. Lange (San Jose State University) and James W. Miller
(Southern Methodist University)
Dana N. Mackenzie (Kenyon College)

1994   C. W. Groetsch (University of Cincinnati)
Dan Kalman (American University)

1995   Anthony P. Ferzola (University of Scranton)
        Paulo Ribenboim (Queen's University)
1996   John H. Ewing (Indiana University, American Mathematical Society)
        James G. Simmonds (University of Virginia)
1997   Colm Mulcahy (Spelman College)
        Lin Tan (West Chester University)
1998   Kevin G. Kirby (Northern Kentucky University)
        Aimée Johnson (Swarthmore College) and Kathleen Madden (Drew University)
1999   David Bleecker (University of Hawaii) and Lawrence J. Wallen (University of Hawaii)
        Aaron Klebanoff (Rose-Hulman Institute of Technology) and John Rickert (Rose-Hulman Institute of Technology)

George Pólya Lecturers
(Mathematical Association of America)

Pólya lecturers speak at six Sectional meetings of the Association, over a two year period.

1991–93   John H. Ewing (Indiana University)
1992–94   Patricia K. Rogers (York University)
1993–95   Carl Pomerance (University of Georgia)
1994–96   Robert Osserman (Stanford University)
1995–97   Underwood Dudley (Depauw University)
1996–98   László Babai (University of Chicago and the Eötvös University)
1997–99   Ronald L. Graham (AT & T Labs)
1998–00   Colin C. Adams (Williams College)
1999–01   Joseph A. Gallian (University of Minnesota, Duluth)

Appendix 10

On Picture-Writing[*]

G. Pólya

Picture-writing … may be the ultimate source of the Greek, Latin, and Gothic alphabets, the letters of which we currently use as mathematical symbols. In what follows, I wish to show how the method of generating functions, important in Combinatory Analysis, can be quite intuitively evolved from "figurate series" the terms of which are pictures (or, more precisely, variables represented by pictures).

Picture-writing is easy to use on paper or blackboard, but it is clumsy and expensive to print. Although I have presented several times the contents of the following pages orally, I hesitated to print it.[†] I am indebted to the editor of the *Monthly* who encouraged me to publish this article.

I shall try to explain the general idea by discussing three particular examples the first of which, although the easiest, will be very broadly treated.

1.1. In how many ways can you change one dollar? Let us generalize the proposed question. Let P_n denote the number of ways of paying the amount of n cents with five kinds of coins: cents, nickels, dimes, quarters and half-dollars. The "way of paying" is determined if, and only if, it is known how many coins of each kind are used. Thus, $P_4 = 1$, $P_5 = 2$, $P_{10} = 4$. It is appropriate to set $P_0 = 1$. The problem stated at the outset requires us

---

[*] Address presented at the meeting of the Association in Athens, GA, March 16, 1956.
[†] I used it, however, in research. See 2, especially p. 156, where the "figurate series" are introduced in a closely related, but somewhat different, form. (Numerals in boldface indicate the references at the end of the paper.)
Excerpted from *The American Mathematical Monthly* 63 (1956), 689–697.

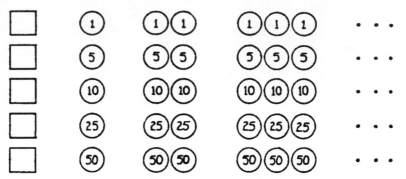

Figure 1. *A complete survey of alternatives.*

to compute P_{100}. More generally, we wish to understand the nature of P_n and eventually devise a procedure for computing P_n.

It may help to visualize the various possibilities. We may use no cent, or just 1 cent, or 2 cents, or 3 cents, or .... These alternatives are schematically pictured in the first line of Figure 1;** "no cent" is represented by a square which may remind us of an empty desk. The second line pictures the alternatives: using no nickel, 1 nickel, 2 nickels, ... The following three lines represent in the same way the possibilities regarding dimes, quarters and half-dollars. We have to choose one picture from the first line, then one picture from the second line, and so on, choosing just one picture from each line; combining (juxtaposing) the five pictures selected, we obtain a manner of paying. Thus, Figure 1 exhibits directly the alternatives regarding each kind of coin and, indirectly, all manners of paying we are concerned with.

The main discovery consists in observing that, in fact, we combine the pictures in Figure 1 according to certain rules of algebra; if we conceive each line of Figure 1 as the *sum* of the pictures contained in it and we consider the *product* of these five (infinite) sums, in short, if we pass from Figure 1 to Figure 2, and we develop the product, the terms of this development will represent the various manners of paying we are concerned with. The one term of the product exhibited in the last line of Figure 2 as an example represents one manner of paying one dollar (putting down no cents, three nickels, one dime, one quarter and one half-dollar). The

---

** A photo of actual coins would be more effective here but too clumsy in the following figures.

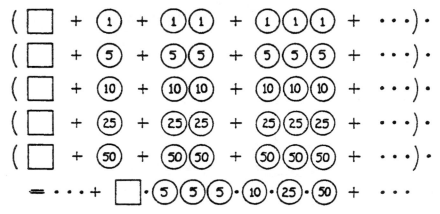

Figure 2. *Genesis of the figurate series.*

sum of all such terms is an infinite series of pictures; each picture exhibits one manner of paying, different terms represent different manners of paying, and the whole series of pictures, appropriately called the *figurate series,* displays all manners of paying that we have to consider when we wish to compute the numbers P_n.

1.2. Yet this way of conceiving Figure 2 raises various difficulties. First, there is a theoretical difficulty: in which sense can we add and multiply pictures? Then, there is a practical difficulty: how can we pick out conveniently from the whole figurate series the terms counted by P_n, that is, those cases in which the sum paid amounts to just n cents?

We avoid the theoretical difficulty if we employ the pictures, these symbols of a primitive writing, as we are used to employing the letters of more civilized alphabets: we regard each picture as the symbol for a variable or *indeterminate.*†

To master the other difficulty, we need one more essential idea: we substitute for each "pictorial" variable (that is, variable represented by a picture) a power of a new variable x, the *exponent* of which is the *joint value of the coins* represented by the picture, as it is shown in detail by Figure

---

† In a formal presentation it may be advisable to restrict the term "picture" to denote a (visible, written, or printed) symbol that stands for an indeterminate; in the present introductory, rather informal, address the word is now and then more loosely used.

Let us pass over two somewhat touchy points: the infinity of variables and the convergence of the series in which they arise. Both are considered in certain advanced theories and both are momentary. They will be eliminated by the next step.

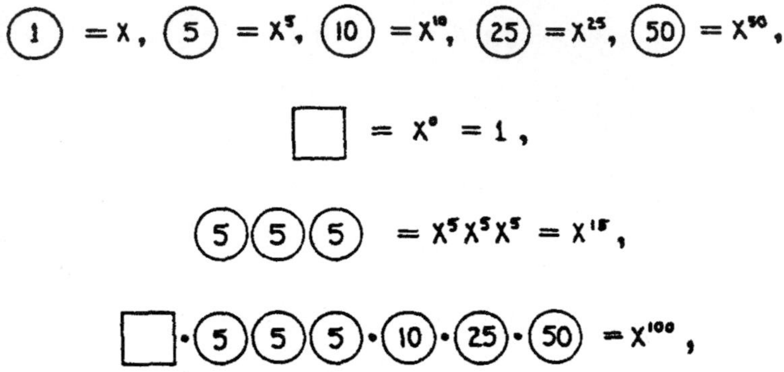

$$\textcircled{1} = x, \quad \textcircled{5} = x^5, \quad \textcircled{10} = x^{10}, \quad \textcircled{25} = x^{25}, \quad \textcircled{50} = x^{50},$$

$$\square = x^0 = 1,$$

$$\textcircled{5}\textcircled{5}\textcircled{5} = x^5 x^5 x^5 = x^{15},$$

$$\square \cdot \textcircled{5}\textcircled{5}\textcircled{5} \cdot \textcircled{10} \cdot \textcircled{25} \cdot \textcircled{50} = x^{100},$$

Figure 3. *Powers of one variable substituted for variables represented by pictures.*

3. The third line of Figure 3 shows a lucky coincidence: we have conceived the three juxtaposed nickels as *one picture*, as the symbol of one variable (corresponding to the use of precisely three nickels). For this variable we have to substitute x^{15} according to our general rule; yet even if we substitute for each of the juxtaposed coins the correct power of x and consider the product of these juxtaposed powers, we arrive at the same final result x^{15}.

The last line of Figure 3 is very important. It shows by an example (see the last line of Fig. 2) how the described substitution affects the general term of the figurate series. Such a term is the product of 5 pictures (pictorial variables). For each factor a power of x is substituted whose exponent is the value in cents of that factor; the exponent of the product, obtained as a sum of 5 exponents, will be the joint value of the factors. And so the substitution indicated by Figure 3 changes each term of the figurate series into a power x^n. As the figurate series represents each manner of paying just once, the exponent n arises precisely P_n times so that (after suitable rearrangement of the terms) the whole figurate series goes over into

$$(1) \qquad P_0 + P_1 x + P_2 x^2 + \cdots + P_n x^n + \cdots.$$

In this series the coefficient of x^n enumerates the different manners of paying the amount of n cents, and so (1) is suitably called the *enumerating series*.

The substitution indicated by Figure 3 changes the first line of Figure 2 into a geometric series:

$$(2) \qquad 1 + x + x^2 + x^3 + \cdots = (1-x)^{-1}.$$

In fact, this substitution changes each of the first five lines of Figure 2 into some geometric series and the equation indicated by Figure 2 goes over into

(3)
$$(1 - x)^{-1}(1 - x^5)^{-1}(1 - x^{10})^{-1}(1 - x^{25})^{-1}(1 - x^{50})^{-1}$$
$$= P_0 + P_1 x + P_2 x^2 + \cdots + P_n x^n + \cdots.$$

We have succeeded in expressing the sum of the enumerating series. This sum is usually termed the *generating function*; in fact, this function, expanded in powers of x, generates the numbers $P_0, P_1, \ldots, P_n, \ldots$, the combinatorial meaning of which was our starting point.

1.3. We have reduced a combinatorial problem to a problem of a different kind: expanding a given function of x in powers of x. In particular, we have reduced our initial problem about changing a dollar to the problem of computing the coefficient of x^{100} in the expansion of the left-hand side of (3). Our main goal was to show how picture-writing can be used for this reduction. Yet let us add a brief indication about the numerical computation.

The left-hand side of (3) is a product of five factors. The well-known expansion of the first factor is shown by (2). We proceed by adjoining successive factors, one at a time. Assume, for example, that we have already obtained the expansion of the product of the first two factors:

$$(1 - x)^{-1}(1 - x^5)^{-1} = a_0 + a_1 x + a_2 x^2 + \cdots,$$

and we wish to go on hence to three factors:

$$(1 - x)^{-1}(1 - x^5)^{-1}(1 - x^{10})^{-1} = b_0 + b_1 x + b_2 x^2 + \cdots.$$

It follows that

$$(b_0 + b_1 x + b_2 x^2 + \cdots)(1 - x^{10}) = a_0 + a_1 x + a_2 x^2 + \cdots.$$

Comparing the coefficient of x^n on both sides, we find that

(4)
$$b_n = b_{n-10} + a_n$$

(set $b_m = 0$ if $m < 0$). By (4), we can conveniently compute the coefficients b_n by recursion if the a_n are already known, and the series (3) can be obtained from (2) in four successive steps each of which is similar to the one we have just discussed.

We add a table that shows the computation of P_{50}. This table exhibits the coefficient of x^n for some values of n in five different expansions. The head of each column shows the value of n, the beginning of each row the last factor taken into account; the bottom row would show P_n for $n = 0, 5, 10, \ldots, 50$ *if* we had computed it. Yet the table registers only the steps needed for computing the answer to our initial question and yields $P_{50} = 50$; that is, one can pay 50 cents in exactly 50 different ways. We leave it to the reader to continue the computation and verify that $P_{100} = 292$;

Table to compute P_{50}

$n = 0$	5	10	15	20	25	30	35	40	45	50
$(1-x)^{-1}$	1	1	1	1	1	1	1	1	1	1	1
$(1-x^5)^{-1}$	1	2	3	4	5	6	7	8	9	10	11
$(1-x^{10})^{-1}$	1	2	4	6	9	12	16		25		36
$(1-x^{25})^{-1}$	1					13					49
$(1-x^{50})^{-1}$	1										50

he can also try to justify the procedure of computation directly without resorting to the enumerating series.*

2.1. Dissect a convex polygon with n sides into $n - 2$ triangles by $n - 3$ diagonals and compute D_n, the number of different dissections of this kind. Examining first the simplest particular cases helps to understand the problem. We easily see that $D_4 = 2$, $D_5 = 5$; of course $D_3 = 1$.

The solution is indicated by the parts (I), (II), and (III) of Figure 4. After the broad discussion of the foregoing solution it should not be difficult to understand the indications of Figure 4.

Part (I) of Figure 4 hints the key idea: we build up the dissections of any polygon that is not a triangle from the dissections of other polygons which have fewer sides. For this purpose, we emphasize one of the sides of the polygon, place it horizontally at the bottom and call it the *base*. One of the triangles into which the polygon is dissected has the base as side; we call this triangle $\triangle$. In the given polygon there are two smaller polygons, one to the left, the other to the right, of $\triangle$. For example, the top line of Figure 4 (I) shows an octagon in which there is a quadrilateral to the left, and a pentagon to the right, of $\triangle$, both suitably dissected. As the figure suggests, we can generate this dissection of the octagon by starting from $\triangle$ and placing on it, from both sides, the two other appropriately pre-dissected polygons. We may hope that building up the dissections in this manner will be useful.

In exploring the prospects of this idea, we may run into an objection: there are cases, such as the one displayed in the second line of Figure 4 (I), in which the partial polygon on a certain side of $\triangle$ does not exist. Yet we can parry this objection: yes, the partial polygon on that side of $\triangle$ (the left side in the case of the figure) *does* exist, but it is degenerate; it is reduced to a mere *segment*.

---

* For the usual method of deriving the generating function, *cf.* **1**, Vol. 1, p. 1, Problem 1.

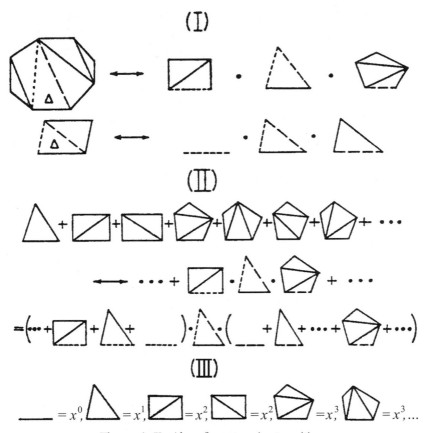

Figure 4. *Key ideas, figurate series, transition.*

Part (II) of Figure 4 shows the genesis of the figurate series. This series, which occupies the first line, is the sum of all possible dissections of polygons with $3, 4, 5, \ldots$ sides. According to Part (I) (as the next line reminds us) each term of the figurate series can be generated by placing two pre-dissected polygons on a triangle $\triangle$, one from the left and one from the right (one or the other of which, or possibly both, may be degenerate). Therefore, as the next line (the last of Figure 4 (II)) indicates, the terms of the figurate series are in one-one correspondence with the expansion of a product of three factors: the middle factor is just a triangle, the other two factors are equal to the figurate series augmented by the segment.

2.2. Part (III) of Figure 4 hints the transition from the figurate series to the enumerating series. Following the pattern set by Figure 3 and

Section 1.2, we substitute for each dissection (more precisely, for the variable represented by that dissection) a power of x the exponent of which is the number of triangles in that dissection. This substitution, indicated by Figure 4 (III), changes the figurate series into

(5) $$D_3x + D_4x^2 + D_5x^3 + \cdots + D_nx^{n-2} + \cdots = E(x),$$

where $E(x)$ stands for enumerating series. The relation displayed by Figure 4 (II) goes over into

(6) $$E(x) = x[1 + E(x)]^2.$$

This is a quadratic equation for $E(x)$ the solution of which is

(7) $$\begin{aligned}
E(x) &= D_3x + D_4x^2 + D_5x^3 + \cdots + D_nx^{n-2} + \cdots \\
&= \frac{1 - 2x - [1 - 4x]^{1/2}}{2x} \\
&= x + 2x^2 + \cdots.
\end{aligned}$$

In fact, to arrive at (7), we have to discard the other solution of the quadratic equation (6) which becomes ∞ for $x = 0$.

2.3. We have reduced our original problem which was to compute D_n to a problem of a different kind: to find the coefficient of x^{n-2} in the expansion of the function (7) in powers of x.* This latter is a routine problem which we need not discuss broadly. We obtain from (7), using the binomial formula and straightforward transformations, that for $n \geq 3$

$$D_n = -\frac{1}{2}\binom{1/2}{n-1}(-4)^{n-1} = \frac{2}{6}\frac{6}{3}\frac{10}{4} \cdots \frac{4n-10}{n-1}.$$

3.1. A (topological) *tree* is a connected system of two kinds of objects, *lines* and *points,* that contains no closed path. A certain point of the tree in which just one line ends is called the *root* of the tree, the line starting from the root the *trunk,* any point different from the root a *knot.* In Figure 5 the root is indicated by an arrow, and each knot by a small circle. Our problem is: *compute T_n, the number of different trees with n knots.*

It makes no difference whether the lines are long or short, straight or curved, drawn on the paper to the left or to the right: only the difference in (topological) connection is relevant. Examining the simplest cases may

---

* For a more usual method *cf.* **3**, Vol. 1, p. 102, Problems 7, 8, and 9.

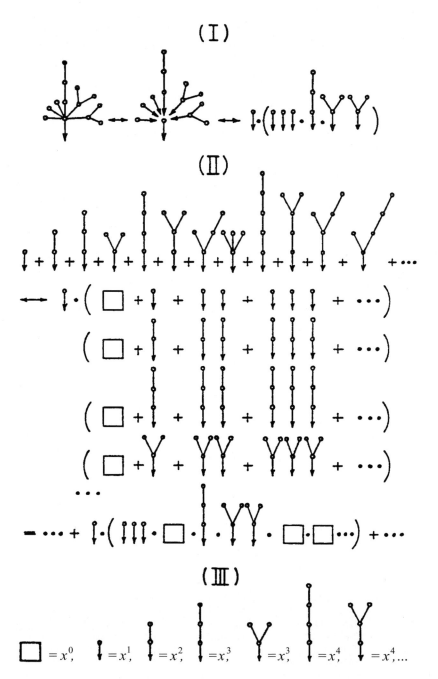

Figure 5. *Key ideas, figurate series, transition.*

help the reader to understand the intended meaning of the problem; it is easily seen that $T_1 = 1$, $T_2 = 1$, $T_3 = 2$, $T_4 = 4$, $T_5 = 9$.[†]

The solution is indicated by the three parts of Figure 5 the general arrangement of which is closely similar to that of Figure 4. The reader should try to understand the solution by merely looking at Figure 5 and observing relevant analogies with all the foregoing figures. He may, however, fall back upon the following brief comments.

The simplest tree consists of root, trunk and just one knot. The key idea is to build up any tree different from the simplest tree from other trees which have fewer knots. For this purpose we conceive, as Figure 5 (I) shows, the "main branches" of any tree as trees (with fewer knots) inserted into the upper endpoint (the only knot) of the trunk. Therefore, as Figure 5 (I) further shows, we can conceive of any tree as the juxtaposition of the simplest tree and of several pictures, each of which consists of one, or two, or more *identical* trees; observe the analogy with the last line of Figure 2.

Part (II) of Figure 5 displays the figurate series: the infinite sum of all different trees. Its genesis is similar to, but more complex than, that of the figurate series of Figure 2. In Figure 2 we see a product of five "virtually geometric" series; in Figure 5 we see a product of an infinity of "virtually geometric" series, multiplied by an initial one term factor (the simplest tree, the common trunk of all trees).

3.2. Part (III) of Figure 5 displays the substitution that changes the figurate series into the enumerating series. By this substitution, each "virtually geometric" series arising in Figure 5 (II) goes over into a proper geometric series the sum of which is known, and the whole relation displayed by Figure 5 (II) goes over into the remarkable relation due to Cayley[*]

$$
\begin{aligned}
&T_1 x + T_2 x^2 + T_3 x^3 + \cdots + T_n x^n + \cdots \\
\text{(8)} \quad &= x(1-x)^{-T_1}(1-x^2)^{-T_2}(1-x^3)^{-T_3}\cdots(1-x^n)^{-T_n}\cdots.
\end{aligned}
$$

[†] The trees here considered should be called more specifically *root-trees*; see **4**, Vol. 11, p. 365. Their definition which is merely hinted here is elaborated in **2**, pp. 181–191; see also the passages there quoted of **5**. It may be, however, sufficient and in some respects even advantageous if, at a first reading, the reader takes the definition "intuitively" and supplements it by examples. Observe that in Cayley's first paper on the subject, **4**, Vol. 3, pp. 242–246, the definition of a tree is not even attempted. Chemistry is one of the sources of the notion "tree": if the points stand for atoms and the connecting lines for valencies, the tree represents a chemical compound.

[*]This form is slightly different from that given in **4**, Vol. 3, pp. 242–246. For other forms see **2**, p. 149.

3.3. By expanding the right-hand side of Equation (8) in powers of x and comparing the coefficient of x^n on both sides, we obtain a recursion formula, that is, an expression for T_n in terms of $T_1, T_2, \ldots, T_{n-1}$ for $n \geq 2$. The reader should work out the first cases and verify by analytical computation the values T_n for $n \leq 5$ which he found before by geometrical experimentation.

References

1. G. Pólya and G. Szegő, *Aufgaben und Lehrsätze aus der Analysis*, 2 volumes, Berlin, 1925.
2. G. Pólya, *Acta Mathematica*, vol. 68 (1937), pp. 145–254.
3. G. Pólya, *Mathematics and Plausible Reasoning*, 2 volumes, Princeton, 1954.
4. A Cayley, *Collected Mathematical Papers*, 13 volumes, Cambridge, 1889–1898.
5. D. König, *Theorie der endlichen und unendlichen Graphen*, Leipzig, 1936.

Appendix II

Generalization, Specialization, Analogy*

G. Pólya

My personal opinion is that the choice of problems and their discussion in class must be, first and foremost, *instructive*. I shall be in a better position to explain the meaning of the word "instructive" after an example. I take as an example the proof of the best known theorem of elementary geometry, the theorem of Pythagoras. The proof on which I shall comment is not new; it is due to Euclid himself (*Elements* V1, 31).

1. We consider a right triangle with sides a, b and c, of which the first, a, is the hypotenuse. We wish to show that

$$(1) \qquad\qquad a^2 = b^2 + c^2.$$

This aim suggests that we describe squares on the three sides of our right triangle. And so we arrive at the not unfamiliar part I of our compound figure. (The reader should draw the parts of this figure as they arise, in order to see it in the making.)

2. Discoveries, even very modest discoveries, need some remark, the recognition of some relation. We can discover the following proof by

* Presented at the summer meeting of the Mathematical Association of America, New Haven, Conn., September 1, 1947. Originally published in the *American Mathematical Monthly* 55 (1948) 241–43.

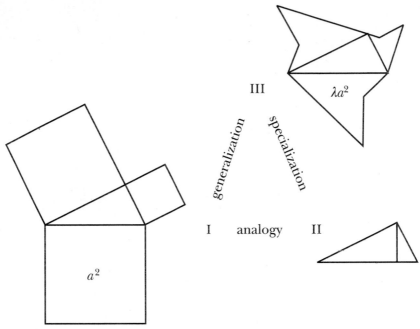

Figure I, II and III

observing the *analogy* between the familiar part I of our compound figure and the scarcely less familiar part II: the same right triangle that arises in I is divided in II into two parts by the altitude perpendicular to the hypotenuse.

3. Perhaps, you fail to perceive the analogy between Figures I and II. This analogy, however, can be made quite explicit by a common *generalization* of I and II which is expressed by Figure III. There we find again the same right triangle, and on its three sides three polygons are described which are similar to each other but arbitrary otherwise.

4. The area of the square described on the hypotenuse in Figure I is a^2. The area of the irregular polygon described on the hypotenuse in Figure III can be put equal to λa^2; the factor λ is determined as the ratio of two given areas. Yet then, it follows from the similarity of the three polygons described on the sides a, b and c of the triangle in Figure III that their areas are equal to λa^2, λb^2 and λc^2, respectively.

Now, if the equation (1) should be true (as stated by the theorem that we wish to prove), then also the following would be true:

(2) $\lambda a^2 = \lambda b^2 + \lambda c^2.$

In fact, very little algebra is needed to derive (2) from (1). Now, (2) represents a *generalization* of the original theorem of Pythagoras: *If three similar polygons are described on the three sides of a right triangle, the one described on the hypotenuse is equal in area to the sum of the two others.*

It is instructive to observe that this generalization is *equivalent* to the special case from which we started. In fact, we can derive the equations (1) and (2) from each other, by multiplying or dividing by λ (which is, as the ratio of two areas, different from 0).

5. The general theorem expressed by (2) is equivalent not only to the special case (1), but to any other special case. Therefore, if any such special case should turn out to be obvious, the general case would be demonstrated.

Now, trying to *specialize* usefully, we look around for a suitable special case. Indeed Figure II represents such a case. In fact, the right triangle described on its own hypotenuse is similar to the two other triangles described on the two legs, as is well known and easy to see. And, obviously, the area of the whole triangle is equal to the sum of its two parts. And so, the theorem of Pythagoras has been proved.

6. 1 took the liberty of presenting the foregoing reasoning so broadly because, in almost all its phases, it is so eminently instructive. A case is instructive if we can learn from it something applicable to other cases, and the more instructive the wider the range of possible applications. Now, from the foregoing example we can learn the use of such fundamental mental operations as generalization, specialization and the perception of analogies. There is perhaps no discovery either in elementary or in advanced mathematics or, for that matter, in any other subject that could do without these operations, especially without analogy.

The foregoing example shows how we can ascend by generalization from a special case, as from the one represented by Figure I, to a more general situation as to that of Figure III, and redescend hence by specialization to an analogous case, as to that of Figure II. It shows also the fact, so usual in mathematics and still so surprising to the beginner, or to the philosopher who takes himself for advanced, that the general case can be logically equivalent to a special case. Our example shows, naively and suggestively, how generalization, specialization and analogy are naturally combined in the effort to attain the desired solution. Observe that only a minimum of preliminary knowledge is needed to understand fully the foregoing reasoning. And then we can really regret that mathematics teachers usually do not emphasize such things and neglect such excellent opportunities to teach their students to think.*

---

* The author's views are presented more fully in his booklet, *How to Solve It* (Princeton, 5th enlarged printing 1948). For more about generalization, specialization and analogy see the sections starting on pp. 97, 164 and 37.

Appendix 12

Heuristic Reasoning in the Theory of Numbers

G. Pólya

A deep but easily understandable problem about prime numbers is used in the following to illustrate the parallelism between the heuristic reasoning of the mathematician and the inductive reasoning of the physicist. The experts may judge whether the parallelism is more serious than the tone of presentation which is adapted to a wider audience.

1. "Till now the mathematicians tried in vain to discover some order in the sequence of the prime numbers and we have every reason to believe that there is some mystery which the human mind shall never penetrate. To convince oneself, one has only to glance at the tables of primes which some people took the trouble to compute beyond a hundred thousand, and one perceives that there is no order and no rule. This is so much more surprising as the arithmetic gives us definite rules with the help of which we can continue the sequence of the primes as far as we please, without noticing, however, the least trace of order."*

So wrote Euler about two centuries ago, yet the prime numbers may inspire the contemporary mathematician with the same feeling of mystery

---

* See L. Euler, *Opera Omnia*, ser. 1, vol. 2, p. 241 or G. Polya, *Mathematics and Plausible Reasoning*, Princeton, vol. 1, p. 91.

This paper originally appeared in the *American Mathematical Monthly* 66 (1959) 375–84.

that Euler so vividly expressed. The primes remain puzzling in spite of many important discoveries made in the meantime. Let us look at some of these discoveries.

The intervals between successive primes are irregular, but these intervals seem to become larger "on the whole" (the primes seem to become scarcer) as we proceed in the sequence of numbers. Since Euler's time a definite law of this phenomenon was discovered (conjectured by Legendre and Gauss, investigated by Chebyshev and Riemann, finally proved by Hadamard and de la Vallée Poussin, proved recently in an essentially different "elementary" manner by Atle Selberg and Paul Erdős). We may formulate this law, the "prime number theorem, intuitively although not quite precisely, as follows: The probability that a large integer x should be a prime, is $1/\log x$ (where $\log x$ is the natural logarithm of x).*

The following short table exhibits the first primes (with two exceptions) classified according to their last digit.

	11		31	41		61	71			101	
3	13	23		43	53		73	83		103	113
7	17		37	47		67			97	107	
	19	29			59		79	89		109	

If we set apart 2 and 5, the prime factors of 10, the last figure in the decimal symbol of a prime cannot be 0, 2, 4, 5, 6, or 8 (since neither 2 nor 5 should be a divisor) and must, therefore, be 1, 3, 7, or 9. Thus, with respect to ten (modulo 10) there are four kinds of primes which are listed in the four horizontal lines of the foregoing table, respectively. Since Euler's time, a general law has been discovered (most of the credit for its discovery is due to Dirichlet) which, applied to our particular case, asserts that there are infinitely many prime numbers of each kind and, what is more, that each kind is equally probable. Therefore, in an extensive

---

* The irregular distribution of primes ("there is no order and no rule") strongly suggests the idea of probability and chance. Yet this is paradoxical: Whether any given integer is a prime or not, can be decided by the "definite rules" of arithmetic—where and how could chance enter the picture? The paradox can be somewhat explained (or deepened) by a physical analogy. The kinetic theory of matter considers the probability distribution of the velocities of the molecules in a gas. Yet this is paradoxical: The velocities resulting from the collision of two molecules can be exactly predicted from the data of the collision by the "definite rules" of classical deterministic mechanics—where and how could chance enter the picture? The determinateness of the simple single event and the probabilistic theory of the highly composite whole may seem to be equally compatible (or incompatible) in both cases.

table of prime numbers there must be roughly as many primes ending with 1 as primes ending with 3.

Euler mentions a table of primes that goes beyond 10^5. Since his time much more extensive tables have been computed, especially in the last decade with the help of machines. Data derived from these tables may suggest problems not yet considered by Euler.

2. The least possible distance between two consecutive primes is 2, if we set apart the unique case of the primes 2 and 3. Two primes having this minimum distance are called *twin primes*. Here is a list of the twin primes under 100:

$$3, 5 \quad 5, 7 \quad 11, 13 \quad 17, 19 \quad 29, 31 \quad 41, 43 \quad 59, 61 \quad 71, 73$$

We can generalize this situation and consider a prime p that is escorted at a given distance d by another prime $p' = p + d$. (This situation is uninteresting unless d is even; we do not care whether there are or are not other primes between p and p'.) Here is a list of all such pairs at the distance 6, in which the first prime does not (but its escort may) exceed 100:

$$5,11 \quad 7,13 \quad 11,17 \quad 13,19 \quad 17,23 \quad 23,29 \quad 31,37 \quad 37,43$$
$$41,47 \quad 47,53 \quad 53,59 \quad 61,67 \quad 67,73 \quad 73,79 \quad 83,89 \quad 97,103$$

It is curious that the second kind of pairs is more numerous. We count 8 pairs of twin primes and exactly twice as many pairs of primes at the distance 6. Let us take now instead of 10^2 the considerably higher bound $3 \cdot 10^7$. Under thirty million there are 152892 primes followed by another prime at the distance 2, but nearly twice as many, namely 304867 primes followed by another at the distance 6.

The numbers of these prime pairs have been obtained by Professor and Mrs. D. H. Lehmer with the use of appropriate computing apparatus; they computed, up to the same limit $3 \cdot 10^7$, the number of primes escorted by another prime at the distance d for $d = 2, 4, 6, 8, \ldots, 70$. I wish to thank them here for their kind permission to use their interesting material. I wish to use some of their results to offer the unprejudiced reader a particularly suitable opportunity for an inductive investigation in pure mathematics.

It will be convenient to introduce here some notation. Let $\pi_d(x)$ stand for the number of those prime numbers p that satisfy two conditions:

$$p \leq x, \quad p + d \text{ is a prime number.}$$

For instance,

$$\pi_2(100) = 8\,, \qquad \pi_2(30\ 000\ 000) = 152892,$$
$$\pi_6(100) = 16, \qquad \pi_6(30\ 000\ 000) = 304867.$$

I set

$$\pi_d(3 \cdot 10^7)/\pi_2(3 \cdot 10^7) = R_d.$$

For instance, $R_6 = 304867/152892 = 1.9940$, approximately. A small part of the material computed by Professor and Mrs. Lehmer is collected in Table I.

d	R_d	18	1.9982	36	1.9997	54	1.9981
2	1.0000	20	1.3311	38	1.0566	56	1.1957
4	0.9979	22	1.1088	40	1.3330	58	1.0349
6	1.9940	24	1.9976	42	2.3987	60	2.6632
8	0.9996	26	1.0910	44	1.1097	62	1.0341
10	1.3317	28	1.1974	46	1.0467	64	0.9999
12	1.9985	30	2.6632	48	1.9965	66	2.2186
14	1.1985	32	0.9970	50	1.3308	68	1.0663
16	1.0001	34	1.0645	52	1.0892	70	1.5977

Table I. Values oF R_d

3. Now, let us start our inductive research. At any moment at which the reader feels inspired, he should interrupt the reading and try to guess the result by himself.

The four kinds of prime numbers that we have considered in Section I (ending with 1, 3, 7 or 9 in the decimal notation, respectively) are known to be equally frequent. Are the 35 kinds of prime numbers with which Table I is concerned also equally frequent? If it were so, all the ratios R_d contained in Table I should be approximately equal to one. In fact, remarkably enough, a few entries in Table I are pretty close to the value 1, but the majority seem to deviate significantly from 1. The analogy with the previous case does not seem to go far. Yet, perhaps, the analogy holds at least in one respect: the ratio $\pi_d(x)/\pi_2(x)$ may converge towards some limit (not necessarily 1) when x tends to infinity, and the ratio $R_d = \pi_d(3 \cdot 10^7)/\pi_2(3 \cdot 10^7)$ entered into Table I may be an approximation to that limit.

We face here a situation somewhat analogous to the situation that the chemists faced around 1800 when they were about to discover the Law of Multiple Proportions. They had to perceive behind their experimental data distorted by unavoidable errors of observation the ratios of simple multiples of the atomic weights, and we have to perceive behind the approximate ratios R_d collected in Table I the true limiting ratios. To guess these limiting ratios is a challenging task.

We have already observed that some values of R_d are very close to 1; they correspond to $d = 2, 4, 8, 16, 32, 64$. (For $d = 2$ the value is exactly 1, but this is trivial.) We can scarcely fail to notice here the powers of 2. By the way, these values of R_d so close to 1 are also the smallest values in the table. Are there other entries in the table so nearly equal to each other?

In trying to answer this question we may notice that the entries corresponding to

$$d = 6, 12, 24, 48$$

are approximately equal to each other, and so are those corresponding to

$$d = 10, 20, 40$$

or those corresponding to

$$d = 14, 28, 56.$$

In general, multiplication of d by 2 seems to leave the value of R_d almost unchanged.

What about multiplication by 3? It approximately doubles the value of R_d in certain transitions, as from

$$2 \text{ to } 6, \quad 4 \text{ to } 12, \quad 8 \text{ to } 24, \quad 16 \text{ to } 48,$$
$$10 \text{ to } 30, \quad 20 \text{ to } 60, \quad 14 \text{ to } 42, \quad 22 \text{ to } 66.$$

Yet it is not so in other cases, as

$$6 \text{ to } 18, \quad 12 \text{ to } 36, \quad 18 \text{ to } 54;$$

in these latter cases the multiplication of d by 3 leaves the value of R_d almost unchanged. How can you account for this different behavior?

And so on, from question to question, by observation and tentative generalization, carefully checking each guess, the reader may discover that many of the values R_d contained in Table I come very close to simple fractions; see Table II.

Table II strongly suggests that R_d *depends only on the decomposition of d into prime factors*. More precisely, just the presence of a prime factor in, or its absence from, the decomposition seems to be relevant; for instance, to all values of d of the form $2^\alpha 3^\beta$ with $\alpha, \beta = 1, 2, 3, \ldots$ there corresponds the same value of R_d (approximately).

d	2	16	6	36	10	14	22	30	42	66	70
	4	32	12	48	20	28	44	60			
	8	64	18	54	40	56					
			24		50						
R_d (approx.)	1		$\dfrac{2}{1}$		$\dfrac{4}{3}$	$\dfrac{6}{5}$	$\dfrac{10}{9}$	$\dfrac{8}{3}$	$\dfrac{12}{5}$	$\dfrac{20}{9}$	$\dfrac{8}{5}$

Table II. Simple Approximations to some R_d

Moreover, to each prime factor of d there seems to correspond a factor of R_d; to the (unavoidable) factor 2 of d, the (trivial) factor 1 of R_d; to the prime factors

$$3, \quad 5, \quad 7, \quad 11$$

of d, the following factors of R_d:

$$\frac{2}{1}, \quad \frac{4}{3}, \quad \frac{6}{5}, \quad \frac{10}{9}$$

respectively. Then, when d is a product of different primes (or powers of different primes) R_d seems to be the product of the corresponding factors.

4. All such observations point to the (conjectural) formula

$$(1) \qquad\qquad \pi_d(x) \sim \pi_2(x)\prod_{p\mid d}\frac{p-1}{p-2},$$

where the product $\prod_{p\mid d}$ is extended over all different *odd* prime factors p of the even number d.* The sign $\sim$ can be interpreted either vaguely or strictly. In a vague interpretation $\sim$ means "approximately equal;" in the strict sense it means "the ratio of the two sides tends to 1 when x tends to ∞." The formula is merely a conjecture which we can conceive quite naively by examining Table I. In Table III, the observed values of R_d, taken from Table I and styled now R_d (obs.), are compared with the corresponding conjectural limiting values, styled R_d (theor.). This comparison yields strong inductive evidence for the conjecture which could be further strengthened by use of other data computed by Professor and Mrs. Lehmer.

---

*The usual abbreviation $a \mid b$ means "a divides b" or "a is a divisor of b. " We shall need later also the abbreviation $a \nmid b$ which means "a is not a divisor of b."

d	R_d (obs.)	R_d (theor.)	24	1.9976	2.0000	48	1.9965	2.0000
2	1.0000	1.0000	26	1.0910	1.0909	50	1.3308	1.3333
4	0.9979	1.0000	28	1.1974	1.2000	52	1.0892	1.0909
6	1.9940	2.0000	30	2.6632	2.6667	54	1.9981	2.0000
8	0.9996	1.0000	32	0.9970	1.0000	56	1.1957	1.2000
10	1.3317	1.3333	34	1.0645	1.0667	58	1.0349	1.0370
12	1.9985	2.0000	36	1.9997	2.0000	60	2.6632	2.6667
14	1.1985	1.2000	38	1.0566	1.0588	62	1.0341	1.0345
16	1.0001	1.0000	40	1.3330	1.3333	64	0.9999	1.0000
18	1.9982	2.0000	42	2.3987	2.4000	66	2.2186	2.2222
20	1.3311	1.3333	44	1.1097	1.1111	68	1.0663	1.0667
22	1.1088	1.1111	46	1.0467	1.0476	70	1.5977	1.6000

Table III. Values of R_d, Observed and "Theoretical"

5. We have before us a precise, general, but enigmatic formula derived from, and quite well verified by, observations. Of course, we wish to understand it, we wish to explain it. When we are looking at it, our situation is similar to that of Newton looking at the laws of Kepler or to that of Niels Bohr looking at Balmer's formula. The word "similar" must be correctly understood. Similar figures may be very different in magnitude, but they show the same proportions, and so do in a sense the three situations we have just compared.

We wish to explain that conjectural formula about prime numbers. Both the irregular distribution of the primes and the structure of the conjectural formula strongly suggest an explanation by probability. I wish to present such an explanation. We shall arrive at it in two steps (of which the second is much more dangerous).

Problem I. *Let p denote a given prime number, d a given integer, and x a large integer chosen at random. Find the probability that neither x nor x + d is divisible by p.*

The reader may visualize the integers as successive intervals of equal length along an infinite straight line, some sort of super-roulette. The interval is red or green, according as the integer is, or is not, divisible by p; among any p consecutive intervals there is always just one that is red. A ball is rolled along the line and steps in the interval x.

We have to distinguish two cases.*

First case: $p \mid d$. In this case $x + d$ falls on a multiple of p (a red space) if, and only if, x itself falls on such a multiple. Therefore, out of any p consecutive numbers (spaces), $p - 1$ are favorable (green) and so the required probability is $(p - 1)/p$.

Second case: $p \nmid d$. Even if x does not fall on a multiple of p, $x + d$ may. Therefore, out of any p consecutive numbers just $p - 2$ are favorable. The required probability is $(p - 2)/p$.

Problem II. *Let d denote a given even integer, and x a large integer chosen at random. Find the probability P_d that both x and $x + d$ are prime numbers.*

In order that both x and $x + d$ should be prime numbers, a sequence of conditions must be satisfied:
First, neither x nor $x + d$ is divisible by 2;
then, neither x or $x + d$ is divisible by 3;
then, neither x nor $x + d$ is divisible by 5;
and so on. The general form of this condition is: neither x nor $x + d$ is divisible by p where p is a prime number.

We have computed above the probability for the fulfillment of any single one of these conditions. Now we have to compute the probability that all these conditions are fulfilled at the same time, all these events are realized simultaneously.

Two difficulties arise here: Are these events independent? How far should we go with p? In fact, these two difficulties may be connected, but at this stage of the game it will be better not to examine them too thoroughly; let us now proceed quickly and see whether anything worthwhile turns up.

Are the events independent? We do not know, but let us assume it. Also the physicist is inclined to assume the independence of the probabilities he deals with—not because he knows that they are independent, but interdependent probabilities are so much more difficult to handle—and so let us assume independence in our case too, although we have no better reasons than the physicist.

Having made this assumption all we have to do is to multiply probabilities computed above. We distinguish three cases:

---

* For the symbols $\mid$ and $\nmid$, see footnote p. 270.

$p = 2$ (which is a divisor of the even number d)

p is odd and is a divisor of d;

p is odd and is not a divisor of d.

Accordingly, the required probability P_d is a product of three kinds of factors:

(2)
$$P_d = \frac{1}{2} \prod_{p|d} \frac{p-1}{p} \prod_{p\nmid d} \frac{p-2}{p}.$$

In this formula (2) (and in the following formulas (3), (4)) the letter p stands for an *odd* prime number.

How far should we go with p? Of course, on the right-hand side of formula (2) we extend the first product over all odd prime factors of the given number d. In the second product, we take all the odd primes not dividing d up to a certain large upper bound, depending on the considered large number x—but let us *postpone* the decision, how far to go, how large that upper bound should precisely be.

We can transform formula (2) as follows

(3)
$$P_d = \prod_{p|d} \frac{p-1}{p-2} \cdot \frac{1}{2} \prod_{p} \frac{p-2}{p};$$

the second product on the right-hand side of (3) is extended over *all* odd primes p under a certain (large, but not yet definitely characterized) upper bound. The first product is extended over the odd prime divisors of d; if d happens to be 2 (or a power of 2) there are no odd prime divisors, that first product is empty, and has to be replaced by 1. Therefore

(4)
$$P_d = \prod_{p|d} \frac{p-1}{p-2} \cdot P_2.$$

Yet the ratio of the probabilities P_d/P_2 should be approximately the same as the ratio of the observed numbers $\pi_d(x)/\pi_2(x)$—and so the formula (4) just derived justifies the conjectural formula (1)—complete success!

6. Unfortunately, our reasoning is vulnerable and the success is illusory. We left a gap in our derivation (we did not decide how far to go with p) and if we try to fill this gap, we run into trouble. The trouble becomes manifest if we try to apply our reasoning to the simplest analogous problem, the result of which is well known.

Problem III. Find the probability that x, a large integer chosen at random, is a prime number.

By reasoning as we did in solving Problem I and assuming the independence of the probabilities involved as we did in solving Problem II we obtain the answer $\prod (p-1)/p$; the product is extended to all primes p not surpassing a certain bound—but what should be the bound? The number x is certainly prime if it is not divisible by any prime $p < x$. This leads to the evaluation of the desired probability

$$(5) \qquad \prod_{p<x} \frac{p-1}{p} \sim \frac{\mu}{\log x}$$

where $\mu = 0.561459\ldots = e^{-c}$ and $c = 0.577215\ldots$ is the familiar constant of Mascheroni and Euler; the asymptotic evaluation in (5) (on the right-hand side of the sign $\sim$) which is valid for $x \to \infty$, is due to Mertens.*

Now, the value (5) is too small. The probability in question is known to be $1/\log x$; this is just the prime number theorem. And we can "explain" somehow why the result is wrong: If the integer x is not divisible by any prime p which does not exceed $x^{1/2}$, x itself must be a prime—and so divisibility by primes exceeding $x^{1/2}$ is, in fact, *not* independent of the smaller primes.

Let us try to modify (5) by considering only primes p not exceeding $x^{1/2}$. This leads us to

$$(6) \qquad \prod_{p \leq x^{1/2}} \frac{p-1}{p} \sim \frac{\mu}{\log(x^{1/2})} = \frac{1.122\ldots}{\log x}$$

(we used Mertens' result (5)) and this value is too large.

Let us, however, imitate the physicists who, without hesitation, modify their theories to fit the observed facts. And so let us do a thing between (5) and (6) and extend the product to all *primes not exceeding x^μ*. We obtain so

$$(7) \qquad \prod_{p<x^\mu} \frac{p-1}{p} \sim \frac{1}{\log x},$$

the right result.

I do not pretend to understand why the introduction of the upper bound x^μ *should* yield the right result. For that matter, when the quanta were introduced, no physicist pretended to understand why energy should be obtainable (as salt or sugar is in the self-service store) only in uniform little packages, in multiples of a certain unit. Yet the criterion of a physical

---

* Cf. G. H. Hardy and E. M. Wright, *An Introduction to the Theory of Numbers*, Oxford, 1938, p. 349, Th. 430.

theory is its applicability. Let us apply the (unintelligible) trick that gave us the right expression for the prime number theorem to our formula (3). Extending the second product to *odd* primes p inferior to x^μ we are led to

$$P_d = \prod_{p|d} \frac{p-1}{p-2} \cdot \frac{1}{2} \prod_{p<x^\mu} \frac{p-2}{p}$$

(8)

$$\sim \prod_{p|d} \frac{p-1}{p-2} \cdot 2 \prod_{p<x^\mu} \frac{(p-2)p}{(p-1)^2} \frac{1}{(\log x)^2};$$

we have used Mertens' result (5). It is easily seen that (8) is equivalent to

(9)
$$P_d \sim 2C_2 \prod_{p|d} \frac{p-1}{p-2} \frac{1}{(\log x)^2},$$

where C_2 stands for the convergent infinite product

$$\prod \left(1 - \frac{1}{(p-1)^2} \right)$$

extended to all odd primes $p=3, 5, 7, 11,\ldots$. The asymptotic formula (9) is due to Hardy and Littlewood, yet even their argument, which is incomparably deeper and more difficult than the one presented here, does not prove (9); it just confers on (9) another kind of plausible evidence. Yet all available numerical data also seem to support (9).

Let us recall that we have attained (9) by combining two analogies, one of which was extremely "natural" and the other (the "trick of the magic μ") extremely "artificial." And let us try to draw the moral: mathematicians and physicists think alike; they are led, and sometimes misled, by the same patterns of plausible reasoning.*

---

* See G. H. Hardy and J. E. Littlewood, Some problems of "Partitio numerorium": On the expression of a number as a sum of primes, *Acta Math.*, vol. 44, 1922, pp. 1–70, especially Conjecture B on p. 42. The more general conjecture on p. 61 (Theorem X 1) is also obtainable by the foregoing reasoning. See also the literature quoted (and criticized) on pp. 32–34, especially the writings of Sylvester, concerning the use of probabilities in questions of similar nature. The crux of the matter may be so expressed: When we consider a fixed number of primes, the "probabilities" introduced can be regarded as "independent," but they cannot be so regarded when the number of primes considered increases in an arbitrary manner. (*Added in proof.* Professor E. M. Wright drew my attention to a paper by the late Lord Cherwell in the *Quart. J. Math.*, vol. 17, 1946, pp. 46–62, which has a certain contact with the present paper, and to a paper by Lord Cherwell and himself which is scheduled to appear in a coming volume of the *Quarterly*.)

Appendix 13

Probabilities in Proofreading

G. Pólya

Two proofreaders, $\mathscr{A}$, and $\mathscr{B}$, read, independently of each other, the proofsheets of the same book. As they finished, A misprints were noticed by $\mathscr{A}$, B misprints by $\mathscr{B}$, C misprints by both, and so, as the result of their joint effort, $A + B - C$ misprints were noticed and corrected. We wish to estimate the number of those misprints that remained unnoticed and uncorrected.

Let M denote the number of all misprints, noticed or unnoticed, in the proofsheets examined, p the probability that proofreader $\mathscr{A}$ notices any given misprint, and q the analogous probability for $\mathscr{B}$. It is an essential assumption that these two probabilities are independent. Hence the expected number of misprints that may be noticed

is:

	by $\mathscr{A}$,	by $\mathscr{B}$,	by both
	Mp,	Mq,	Mpq,

respectively.

In order to arrive at the desired estimate we assume that the expected numbers are approximately equal to the numbers actually found, in symbols

$$Mp \sim A, \quad Mq \sim B, \quad Mpq \sim C$$

and so

This paper originally appeared in the *American Mathematical Monthly* 83 (1976) 42.

$$M = \frac{Mp \cdot Mq}{Mpq} \sim \frac{AB}{C}.$$

Hence the number of misprints that remained unnoticed is

$$= M - (A + B - C) \sim \frac{AB}{C} - (A + B - C) = \frac{(A - C)(B - C)}{C}.$$

This is the desired estimate.

Appendix 14

Cast of Characters— A Glossary of Names

A[braham] Adrian Albert (1905–1972) Albert was born in Chicago and educated there. He became a student of Dickson's at the University of Chicago after which he taught briefly at Princeton and Columbia before returning to Chicago for most of his career. He was an algebraist.

James Waddell Alexander (1888–1971) Born in New Jersey, educated at Princeton University, Alexander taught at Princeton from 1911 until 1933 when he joined the faculty of the Institute for Advanced Study. His field was topology, in particular, knot theory.

Emil Artin (1898–1962) Born in Hamburg, Artin studied in Leipzig. After a postdoctoral year in Göttingen, he joined the faculty at Hamburg till he was forced to leave in 1937. He then taught at Indiana and Princeton till the end of the war made it possible for him to return to Hamburg. He was a ring theorist. He solved Hilbert's 17th problem from the set of 23 outlined at the Congress in Paris in 1900.

Harold Maile Bacon (1907–1992) Born in Los Angeles, Bacon received his education beyond secondary school entirely at Stanford University, where he wrote his Ph.D. dissertation under the direction of Harald Bohr (though when Bohr returned to Copenhagen, final approval was given by J. V. Uspensky). He spent essentially his entire professional career at Stanford. He was a legendary teacher who in later years became concerned with teacher education.

Stefan Bergman (1895–1977) Bergman studied at the University of Berlin and later held a position there until he was forced to leave in 1933. He went first to Russia, then to Paris, finally coming to the United States in 1939 where he held positions at Brown and Harvard before going to Stanford. He is best known for the Bergman kernel.

Paul Isaac Bernays (1888–1977) Born in London of Swiss parents, Bernays studied at Berlin and Göttingen where he wrote his dissertation under the direction of Landau in analytic number theory. He taught at the University of Zürich, Göttingen and the Eidgenössische Technische Hochschule in Zürich. He worked with Hilbert on foundations and influenced Lakatos in his work on heuristics. Bernays shared this interest in heuristics with Pólya.

Ludwig Georg Elias Moses Bieberbach (1886–1982) Bieberbach was born in Goddelau, Germany. He became a faculty member at the University of Berlin and is remembered mainly for his famous conjecture on schlicht functions, finally resolved by de Branges in 1984. Bieberbach was an active Nazi during World War II and founded the infamous journal, *Deutsche Mathematik,* designed to promote mathematics in the "German style".

George David Birkhoff (1884–1944) Born in Michigan, Birkhoff studied at the University of Chicago and at Harvard, after which he taught at Wisconsin, Princeton and finally Harvard (1912). For many years he was regarded as the dean of American mathematics. He worked in differential equations, dynamical systems, ergodic theory, among other areas.

Wilhelm Johann Eugen Blaschke (1885–1962) Born in Graz, Blaschke studied at Göttingen with Klein, Hilbert and Runge, then in Bonn with Study. He held positions in Prague, Leipzig, Königsberg and Hamburg. His work was in integral geometry and functions of several complex variables. He, like Bieberbach, was an active Nazi during World War II.

Hans Frederik Blichfeldt (1873–1945) Born in Denmark, Blichfeldt spent his early years in the lumber industry in the Pacific Northwest before becoming a student at Stanford University. After graduation he became a student of Sophus Lie at Leipzig, where he received his Ph.D. in 1898. He spent most of his career at Stanford, working in group theory and the geometry of numbers.

Harald August Bohr (1887–1951) Bohr was the younger brother of the physicist Niels Bohr and was known not only as a mathematician but as an extraordinarily gifted soccer player. He studied mathematics at the University of Copenhagen where he remained throughout his career. He was an analytic number theorist and spent much of his life studying problems related to the Riemann Hypothesis.

János Bolyai (1802–1860) Born in Kolozsvár, Hungary (now Cluj, Romania), Bolyai was the son of the mathematician Farkas (Wolfgang)

Bolyai. Along with Gauss and Nikolai Lobachevsky, he is credited with the discovery of noneuclidean geometry.

[Felix Édouard Justin] Émile Borel (1871–1956) Born in St. Affrique, France, Borel was educated at the École Normale Supérieure. He taught first at the University of Lille and finally at the Sorbonne (1909), after which he became director of the Institut Henri Poincaré. His work was in analysis.

William Burnside (1852–1927) Burnside was born in London. He studied at Cambridge and received his doctorate from Dublin. He then served as professor of mathematics at the Royal Naval College in Greenwich. Burnside is best known for his work in group theory but he also worked in elliptic functions, probability theory, among others fields.

Constantin Carathéodory (1873–1950) Born to a Greek family in Berlin, Carathéodory studied in Berlin and Göttingen, where he worked with Minkowski. After some years in Germany he returned to Greece to join the faculty of the university at Smyrna. From there he moved to Athens and in 1924 to the University of Munich where he spent the remainder of his career. His interests were in analysis, in particular the calculus of variations.

Élie Joseph Cartan (1869–1951) Cartan was born in Dolomieu in the French Alps. After studying at the École Normale Supérieure he taught at the universities of Montpellier, Lyon, Nancy and Paris. He worked in Lie algebras and a variety of other fields. His son, Henri, is also a mathematician of note.

Mary Lucy Cartwright (1900–1998) Born in Ayno, England, Cartwright was educated at Oxford and became a lecturer in mathematics at Cambridge in 1935. She worked with Littlewood in the theory of functions and in 1969 she was named Dame Commander of the British Empire.

Sarvandaman Chowla (1907–1995) Born in London, educated in Punjab and Cambridge, Chowla served on the faculty at the University of Audhra (India) and the University of Kansas before going to the University of Colorado where he spent the remainder of his career. He was a number theorist.

Richard Courant (1888–1972) Born in Lublinitz in Poland, Courant studied at Breslau (now Wrocław), Zürich, and, finally, at Göttingen under the direction of Hilbert. He went to teach at the University of Muenster, returning to Göttingen to head up the Mathematical Institute there. He was forced to leave in 1934, whence he founded what is today the Courant Institute at New York University (modeled on the

Institute at Göttingen). He worked in applied analysis. He married Nina Runge, Carl Runge's daughter.

George Bernard Dantzig (1914–) Dantzig was born in Portland, Oregon. He was educated at the universities of Maryland and Michigan, receiving his Ph. D. from the University of California, Berkeley, in 1946. He is professor of operations research and computer science at Stanford University, now emeritus. He is a pioneer in operations research and linear programming.

Harold Davenport (1907–1969) Davenport was born in Accrington, England, and studied at the University of Manchester and Trinity College, Cambridge, after which he taught at Manchester; Bangor; University College, London; and finally at Trinity College, Cambridge. He was a number theorist.

Nicolaas Govert de Bruijn (1918-) Born in The Hague, de Bruijn studied at Leyden and Amsterdam. He has held positions in Delft, Amsterdam and Eindhoven. His research has been concentrated on analytic number theory and combinatorics.

[Julius Wilhelm] Richard Dedekind (1831-1916) Dedekind was born in Brunswick. He studied at the University of Göttingen and worked with Riemann and Dirichlet. After teaching at Göttingen, he moved to the Eidgenössische Technische Hochschule in Zürich in 1858, finally returning to Brunswick in 1862. His contributions were largely in analysis and number theory.

Max Wilhelm Dehn (1878-1952) Dehn was born in Hamburg. He studied at Göttingen before becoming a professor at Frankfurt. Forced to leave Germany in 1935, he moved to the University of Idaho, Southern Branch (now Idaho State University), thence to the Illinois Institute of Technology, St. John's College, Annapolis, and eventually Black Mountain College in North Carolina. He solved the third problem from Hilbert's famous list from the Paris Congress of 1900.

Leonard Eugene Dickson (1874-1954) Born in Iowa, Dickson was educated in Iowa and at the universities of Texas, Chicago, Leipzig and Paris before returning to the United States to teach at Texas. Most of his career was spent at the University of Chicago. He was a number theorist and algebraist.

Arthur Erdélyi (1908-1977) Born in Budapest, Erdélyi studied in Brno and Prague, where he received his Ph.D. in 1938. He taught at Edinburgh and the California Institute of Technology. His research was in differential equations and special functions.

Paul Erdős (1913–1996) Erdős was born in Budapest and educated at the university there. He held almost no professional appointments during his long career (one year at Purdue University) but instead traveled around the world visiting mathematicians and collaborating with them on research. He was one of the most colorful figures in the history of mathematics. His fields were number theory and combinatorics.

Leonhard Euler (1707–1783) Born in Basel, Euler at the age of 16 obtained his doctorate. Three years later he moved to the Academy of Sciences in St. Petersburg where he spent most of his career, though he spent 25 years in Berlin in the court of Frederick the Great. He did major work in analysis, geometry, celestial mechanics, number theory, combinatorics, and what we would now call graph theory, as well as various applied areas like naval architecture. He was one of the most prolific of mathematicians; his collected works run to something like 80 folio volumes.

Lipót Fejér (1880–1959) Born in Pécs, Hungary, and educated at the universities of Budapest and Berlin, Fejér later taught at Kolozsvár (now Cluj, Romania) and Budapest, where he remained from 1911 till his death. His contributions were primarily in harmonic analysis, conformal mapping and entire functions.

Miháhy Fekete (1886–1957) Fekete was born in Zenta, Yugoslavia. He studied at the University of Budapest and then at Göttingen. He taught in Belgrade and then, from 1928, at the University of Jerusalem, where he remained until his death. He worked in applied analysis and is remembered for important work in capacity.

Maurice René Fréchet (1878–1973) Born in Maligny, France, and a student at the École Normale Supérieure, Fréchet held appointments at the universities in Poitiers, Strasbourg, and Paris, contributing to probability theory, statistics, topology and analysis.

[Karl] Rudolf Fueter (1880–1950) Born in Basel, Fueter studied there and in Göttingen, Paris, Vienna and London, after which he taught in Basel, Karlsruhe, and the University of Zürich. He was the principal organizer of the International Congress in Zürich in 1932. Fueter worked in number theory and the theory of functions.

Karl Friedrich Geiser (1843–1934) Geiser was born in Switzerland and educated at the University of Berlin, where he taught before returning to the Eidgenössische Technische Hochschule in Zürich. He organized the first International Congress of Mathematicians, held in Zürich in 1897. His own research was in algebraic geometry.

Ferdinand Gonseth (1890–1975) Gonseth studied at the Eidgenössische Technische Hochschule (ETH) in Zürich, a student of Kollros. After appointments at the University of Zürich and the University of Bern, he returned to the ETH. Though his dissertation was in projective geometry, his work was much broader, ranging from mathematical physics to the foundations of mathematics.

Alfréd Haar (1885–1933) Haar was born in Budapest. He studied at Göttingen where he took his doctorate under Hilbert in 1909. After teaching briefly in Göttingen, he returned to Hungary to teach at Kolozsvár (now Cluj, Romania), Budapest and Szeged. He worked in analysis and group theory and is remembered best for Haar measure.

Jacques Salomon Hadamard (1865–1963) Born in Versailles, France, after studies at the École Normale Supérieure, he held positions at the Lycée Buffon in Paris, the University of Bordeaux, the Sorbonne, the Collège de France, the École Polytechnique and the École Centrale des Arts et Manufactures. He worked in analysis, number theory and mathematical physics.

Philip Hall (1904–1982) Born in Hampstead, England, Hall was educated at King's College, Cambridge, where he was elected a fellow in 1927, a position he held for the rest of his life. He is known primarily for his work in group theory.

Frank Harary (1921–) Harary was born in New York. He was educated at Brooklyn College and the University of California, Berkeley, where he received his Ph.D. in 1948. He has taught at Michigan (Ann Arbor) and New Mexico State University. His work is principally in graph theory.

Godfrey Harold Hardy (1877–1947) Born in Cranleigh, England, Hardy was admitted to Trinity College, Cambridge in 1896. He studied and taught there till he moved to Oxford in 1920, but he returned to Cambridge in 1931. He was generally regarded as the preeminent English mathematician of the first half of the twentieth century. His work was in analysis and number theory.

Erich Hecke (1887–1947) Born in Buck-Posen, Hecke studied at Breslau (now Wrocław, Poland), Berlin and Göttingen where he became an assistant to Klein and Hilbert. He held appointments at Basel and Göttingen before moving to Hamburg in 1919. He worked in number theory.

John Herriot (1916–) Herriot was born in Winnipeg, Manitoba, and educated at Manitoba and Brown University, where he received his Ph.D. in 1938. After two years at Brown and a year at Yale, he moved to Stanford in 1942 where he is a professor of computer science emeritus.

Joseph Hersch (1925–) Hersch was born in Geneva and became a student at the Eidgenössische Technische Hochschule in Zürich. Since 1948 he has been a faculty member there. He works in applied analysis.

David Hilbert (1862–1943) Born in Königsberg (Kaliningrad) and educated there and at Heidelberg, after additional studies in Leipzig and Paris, Hilbert returned to Königsberg. He moved to Göttingen in 1895 where he remained till his retirement in 1930. He was widely viewed as the most important mathematician of his generation. He contributed to analysis, the foundations of mathematics, number theory, and integral equations.

Heinz Hopf (1894–1971) Hopf was born in Breslau (now Wrocław, Poland). He studied at Heidelberg, Berlin and Göttingen. After a year at Princeton he took the professorship vacated by Hermann Weyl at the Eidgenössische Technische Hochschule in Zürich in 1931. From 1955 to 1958 he was president of the International Mathematical Union. He was a topologist.

Adolf Hurwitz (1859–1919) Hurwitz was born in Hildesheim, Germany. He was educated at the universities of Munich, Berlin, Göttingen and Leipzig where his dissertation was directed by Klein. He taught at the University of Königsberg before going to the Eidgenössische Technische Hochschule in Zürich in 1892, where he replaced Frobenius. He remained in Zürich for the remainder of his career. His principal contributions were made in complex analysis and algebra.

Albert Edward Ingham (1900–1967) Ingham was born in Northampton, England. He studied at Trinity College, Cambridge and taught briefly at Leeds before returning to Cambridge where he remained. He was an analytic number theorist who worked on the Riemann Hypothesis.

Marc Kac (1914–1984) Born in Drzemieniec, Poland, Kac was educated at the John Kasimir University in Lvov. After teaching there and at The Johns Hopkins University and Cornell, he moved to the Rockefeller University in New York. He worked in a variety of fields, including probability theory, mathematical physics, and number theory.

Béla von Kerékjártó (1898–1946) Kerékjártó was born in Budapest and studied there, receiving his Ph.D. from the University of Budapest. He taught in Szeged beginning in 1922 and in Budapest beginning in 1938. His field was topology.

[Christian] Felix Klein (1849–1925) Klein was born in Düsseldorf. He studied at the University of Bonn and taught at Erlangen, Munich and Leipzig before joining the faculty in Göttingen in 1886. He was largely responsible for the preeminence of Göttingen at the turn of the century, having brought Hilbert there from Königsberg in 1895. Klein's work was in geometry and he is remembered for his Erlanger Programm relating geometry to groups.

Konrad Hermann Theodor Knopp (1882–1957) Knopp was born in Berlin and studied in Lausanne and Berlin. He taught first at the University of Nagasaki and at the German Chinese Academy in Tsingtao, China. Returning to Germany he held successively posts at Berlin, Königsberg and Tübingen. He worked in analysis, complex functions, and series in particular.

Donald Ervin Knuth (1938–) Knuth was born in Milwaukee, Wisconsin. He was educated at Case Institute of Technology (now Case Western Reserve University) and the California Institute of Technology. He taught at the latter till he moved to Stanford University in 1968. He is a—some would say the—preeminent contributor to the development of computer science.

Dénes Kőnig (1884–1944) Kőnig was born in Budapest and was educated at the university there and at Göttingen. He taught at the Technical University in Budapest. While best known for his work in graph theory, he also worked in topology and geometry.

Gyula (Julius) Kőnig (1849–1913) Born in Györ, Hungary, Kőnig studied in Vienna (medicine), Berlin (mathematics and philosophy) and finally Heidelberg (where he studied all of these). He received his doctorate in 1870. He became a professor at the Technical University of Budapest in 1874. He worked mainly in algebra, but also in analysis and, late in his life, in set theory.

József Kürschák (1864–1933) Kürschák spent his whole career in Budapest, first as a student, then as a professor at the Technical University. He worked in the theory of valuations.

Joseph Louis Lagrange (1736–1813) Lagrange was born in Turin (now Italy, then Savoie). He succeeded Euler at the Berlin Academy of Sciences, where he served between 1766 and 1786. He then returned to Paris to positions at the École Normale Supérieure and the École Polytechnique. Like Euler he worked in practically every branch of mathematics known in his time. He was a giant in the history of mathematics.

Imre Lakatos (1922–1974) Lakatos was born in Hungary. He studied at the University of Debrecen and Cambridge, where he wrote his thesis on a topic suggested by Pólya. This later became his classic book *Proofs and Refutations*, a complex Socratic dialogue on the Euler-Descartes formula for polyhedra. Lakatos was a professor at the London School of Economics.

Edmund Georg Hermann Landau (1877–1938) Born in Berlin, Landau was educated there. He received his doctorate from the University of Berlin where he wrote a dissertation under Frobenius. He replaced Minkowski at Göttingen and remained there till 1933 when he lost his position due to the rise of the Nazis. His work was in analytic number theory.

Pierre Simon de Laplace (1749–1827) Born at Beaumont-en-Auge, Normandy, Laplace received his education in Beaumont and Caen. He went on to teach at the École Militaire and the École Normale Supérieure. He served as interior minister under Napoleon. His accomplishments in mathematics, astronomy, and physics are numerous, but especially outstanding were his pioneering results in probability theory and celestial mechanics.

Henri Léon Lebesgue (1875–1941) Lebesgue was born in Beauvais, France. He studied at the École Normale Supérieure. After teaching at the Lycée in Nancy and the universities of Rennes, Poitiers, and at the Sorbonne, he was named to the Collège de France in 1921. His name most often arises in the context of Lebesgue integration.

Derrick Henry Lehmer (1905–1991) Born in Berkeley, California, the son of a number theorist, D. N. Lehmer, D. H. Lehmer was educated at Berkeley and received his degrees from the University of California, Berkeley and Brown University, where he wrote his dissertation under the direction of J. D. Tamarkin. He taught at Lehigh University before joining the Berkeley faculty. His fields were number theory and computing.

Paul Pierre Lévy (1886–1971) Lévy was born in Paris. He was educated at the École Polytechnique and the École Nationale Supérieure des Mines and later taught at both. His field was probability theory, though he contributed to functional analysis and differential equations as well.

James Edensor Littlewood (1885–1977) Littlewood was born in Rochester, England. He studied at Trinity College, Cambridge, and spent his whole career there except for three years at the University of Manchester. He collaborated with Hardy for 35 years, working on the theory of functions, inequalities and analytic number theory.

Charles (Karl) Loewner (1893–1968) Loewner was born in Lauy, Bohemia. He was educated in Prague where he received his Ph.D. from Charles University in 1917. He taught at Prague, Berlin, Brown, and Syracuse (N.Y.) before joining the Stanford faculty. His work was in analysis; he made a major contribution to the solution of the Bieberbach conjecture.

Franz Mertens (1840–1927) Mertens was born in Prussia. He studied in Berlin with Kronecker and Kummer, then held appointments at Cracow and Graz before moving to Vienna in 1894. He worked in algebraic geometry and analytic number theory.

Hermann Minkowski (1864–1909) Born in Alexota, Russia (now Kaunas, Lithuania), Minkowski was raised in Königsberg where he did his principal university work. He also studied at the University of Berlin. He taught at Bonn, Königsberg, the Eidgenössische Technische Hochschule in Zürich, and finally Göttingen. He is remembered for his work in the geometry of numbers and for his ideas on the space-time continuum, work that influenced Einstein.

[Magnus] Gösta Mittag-Leffler (1846–1927) Mittag-Leffler was born in Stockholm and studied there and with Weierstraß in Berlin. After appointments at the universities of Uppsala and Helsinki, he returned to Stockholm. He worked in function theory. He built one of the great mathematical libraries in the world and his name is perpetuated in the name of the Institut Mittag-Leffler in Djursholm, Sweden.

Rolf Herman Nevanlinna (1895–1980) Nevanlinna was born in Finland and spent his whole professional life, as student and faculty member, at the University of Helsinki. He developed harmonic measure and Nevanlinna theory. The Nevanlinna Prize is given every four years at the International Congresses for outstanding work in information science. Between 1959 and 1962 he was president of the International Mathematical Union.

Jerzy Neyman (1894–1981) Born in Bendery (Moldavia), Russia, (later, for a time, part of Romania), Neyman's family was Polish. He studied at the University of Kharkov and worked in England until he came to the United States in 1938, where he became a faculty member at the University of California, a position he occupied till the end of his life. His work is in statistics.

[Amalie] Emmy Noether (1882–1935) The daughter of the mathematician Max Noether at the University of Erlangen, she was educated there and in Göttingen. She taught at Göttingen without holding an official appointment since women were not given regular positions there at the time. In 1935 she came to the United States where she worked at Bryn Mawr and Princeton. She did outstanding work in algebra.

William Fogg Osgood (1864–1943) Born in Boston, Osgood was both a student and a teacher at Harvard. He also studied at Göttingen, where he was a student of Klein's, and at Erlangen. In 1933 he left Harvard and taught for two years at the National University of Beijing. He worked in differential equations, function theory and the calculus of variations.

Albert Pfluger (1907–1993) Pfluger was born in Oensingen, Switzerland, studied at local schools and attended the Eidgenössische Technische Hochschule in Zürich, where he received his doctorate in 1935 under the direction of Pólya. After teaching for three years in secondary schools, he received an appointment in 1938 at the University of Fribourg. Five years later he joined the faculty of the Eidgenössische Technische Hochschule and remained there for the remainder of his career. He worked in complex analysis.

[Charles] Émile Picard (1856–1941) Picard was born in Paris. He earned his doctorate at the École Normale Supérieure and held positions at the University of Toulouse, the Sorbonne and the École Normale Supérieure, as well as positions with the Académie des Sciences and the Bureau des Longitudes. He worked in various branches of analysis and algebraic geometry but is largely remembered for his work in differential equations. He married the daughter of Charles Hermite.

Michel Plancherel (1885–1967) Plancherel was born in Bussy, Switzerland, and studied in Fribourg (Switzerland), Göttingen and Paris. He taught mathematics at the universities of Geneva, Fribourg and finally, beginning in 1920, at the Eidgenössische Technische Hochschule in Zürich. His principal contributions were in analysis.

[Jules] Henri Poincaré (1854–1912) Born in Nancy, Poincaré studied at the École Polytechnique and at the École des Mines. He first taught at the University of Caen but went to the University of Paris in 1881 and stayed there till 1912. He was the most influential French mathematician of his day, working in a wide variety of fields: analysis situs (topology), number theory, algebraic geometry, complex variables and a number of applied fields.

Louis Poinsot (1777–1859) Poinsot was born in Paris. He studied at the École Polytechnique and the École des Ponts et Chaussées and later held positions at the École Polytechnique and the Université Impériale. He worked in geometry and applied mathematics.

Siméon Denis Poisson (1781–1840) Born in Sceaux, France, Poisson studied medicine before deciding to study mathematics at the École Polytechnique, where he later taught, before moving on to the Bureau des Longitudes and eventually to the University of Paris. He worked in probability, analysis and various applied fields.

Alfred Pringsheim (1850–1941) Pringsheim was born in Ohlau, Silesia. He studied in Berlin and taught at the University of Munich, where he worked in real and complex functions. His son-in-law was the novelist, Thomas Mann.

Tibor Radó (1895–1965) Radó was born in Budapest. He studied in Szeged under F. Riesz. In the United States he held posts at Harvard, Rice, and from 1930 at Ohio State. His work was mainly in various branches of analysis, as well as in topology.

Srinivasa Aiyangar Ramanujan (1887–1920) Born in Erode (near Madras), India, Ramanujan in his formative years was largely self-taught until in 1913 he contacted G. H. Hardy at Cambridge and sent him some of his remarkable discoveries. Hardy was so impressed he arranged for Ramanujan to come to Cambridge where they collaborated until 1919. Seriously ill, Ramanujan returned to India where he died the following year, leaving a legacy of incredibly deep unproved and proved theorems and formulas in analysis and number theory.

Alfréd Rényi (1921–1976) Rényi was born in Budapest. He studied in Szeged with F. Riesz and worked in analytic number theory, but he is best known for his work in probability. He is remembered too as a raconteur and commented once that a mathematician is a machine for changing coffee into theorems, to which Turán added that weak coffee produces only lemmas.

Georg Friedrich Bernhard Riemann (1826–1866) Riemann was born in Breselenz, Hanover. He studied at Göttingen and Berlin and received his doctorate at Göttingen in 1851. He worked closely with Dirichlet and replaced him at Göttingen when Dirichlet died in 1859. Riemann contracted tuberculosis in 1862 and died in Italy four years later. He left behind an amazing legacy of brilliant mathematical insights.

Frigyes (Frederick) Riesz (1880–1956) Riesz was born in Györ, Hungary. He studied at the Eidgenössische Technische Hochschule in Zürich, in Budapest, Göttingen and Paris. After an appointment at the University of Kolozsvár (now Cluj, Romania), he moved to Szeged in 1920 where he founded the János Bolyai Mathematical Institute and the *Acta scientiarum mathematicarum*. In 1946 he returned to the University of Budapest. He did pioneering work in functional analysis and operator theory.

Marcel Riesz (1886–1969) The younger brother of F. Riesz, Marcel Riesz was born in Györ and studied in Budapest. He moved to Sweden in 1908, where he spent the rest of his life, first at the University of Stockholm (1911), then at the University of Lund (1926). He worked in functional analysis, partial differential equations and algebra.

Paul C. Rosenbloom (1920–) Rosenbloom was born in Portsmouth, Virginia. He was educated at the University of Pennsylvania and Stanford where he received his Ph.D. in 1944. He taught at Brown, Syracuse (NY), and Minnesota (Twin Cities) before joining the faculty of Teacher's College, Columbia. He worked in function theory and differential equations prior to moving to mathematics education.

Carl David Tolmé Runge (1856–1927) Runge was born in Bremen. He was a student of Weierstraß and Kronecker and became a professor at the Polytechnic School in Hanover in 1886. In 1904 he moved to Göttingen where he held a chair in applied mathematics.

Max Menahem Schiffer (1911–1997) Born in Berlin, Schiffer was educated there and at the Hebrew University in Jerusalem. He joined the Stanford faculty in 1951 after various appointments at the Hebrew University, Harvard, and Princeton. He worked in classical analysis and applied mathematics.

Ludwig Schlesinger (1864–1933) Schlesinger was born in Nagyszombat, Hungary (now Trnava or Tyrnau, Slovakia). He studied in Heidelberg and Berlin where he received his doctorate in 1887. He taught at the University of Kolozsvár (now Cluj, Romania) and very briefly at the University of Budapest, from which he moved to Giessen in 1911, where he stayed till his retirement in 1930. He worked in automorphic functions and differential equations.

Isaac Jacob Schoenberg (1903–1990) Schoenberg was born in Galatz, Romania. He was educated at the University of Jassy in Moldavia, and received his Ph.D. there in 1926. After a series of temporary appointments he taught at Swarthmore College and the universities of Pennsylvania and Wisconsin. He is known principally for his work on splines. Schoenberg chose relatives well—his first wife was the daughter of Landau; his brother-in-law was Hans Rademacher.

Issai Schur (1875–1941) Born in Mogilyov, Russia, Schur studied in Berlin, receiving his Ph.D. there in 1901. He then taught in Bonn before returning to the University of Berlin in 1916. He had to leave Germany in 1935 and went to what was then Palestine. He worked in several fields but is best known for his work in representation theory of groups.

[Karl] Hermann Amandus Schwarz (1843–1921) Born in Hermsdorf, Poland (now part of Germany), Schwarz studied in Berlin (chemistry as well as mathematics). After appointments at Halle, the Eidgenössische Technische Hochschule in Zürich and, finally, Göttingen, he succeeded Weierstraß at Berlin. He worked in analysis; most students of elementary mathematics will recognize the name from the Schwarz inequality for integrals. His wife was the daughter of E. E. Kummer.

Wacław Sierpiński (1882–1969) Sierpiński was born and educated in Warsaw but took his doctorate at the Jagiellonian University in Cracow. From there he went to the University of Lvov, thence back to Warsaw. His work was in set theory and number theory.

Jakob Steiner (1796–1863) Born in Utzentorf, Switzerland, into a peasant family, Steiner studied at a Pestalozzi School in Yverdon, followed by the universities in Heidelberg and Berlin. He joined the faculty in Berlin in 1834. His contributions were mainly in geometry.

Ernst Steinitz (1871–1928) Steinitz was born in Laurahütte, Silesia, and studied in Breslau (now Wrocław, Poland) and Berlin, after which he taught in Berlin-Charlottenberg and Breslau before becoming a professor at Kiel in 1910. He was an algebraist.

Ottó Szász (1884–1952) Szász was born in Hungary and studied in Budapest where he did his doctoral work. He taught at Frankfurt but had to leave for the United States in 1933. He taught at the University of Cincinnati for the remainder of his career.

Gábor Szegő (1895–1985) Born in Kunhegyes, Hungary, Szegő studied in Budapest, Berlin and Göttingen though he was awarded his doctorate from the University of Vienna. He succeeded Knopp at the University of Königsberg and remained there till forced to leave in 1934. He moved to Washington University, St. Louis, and, in 1938, to Stanford. His best known research was in orthogonal polynomials.

Jacob Daniel Tamarkin (1888–1945) Tamarkin was born in Chernigoff, Russia. He studied at the University of St. Petersburg, where he taught until he came to Dartmouth College in 1925. From 1927 until his retirement he taught at Brown University. He worked in analysis and applied mathematics.

Otto Toeplitz (1881–1940) Born in Breslau (now Wrocław, Poland), he studied there and in Göttingen. He later taught at Kiel and Bonn, but in 1938 he was forced to leave. He went to the Hebrew University in Jerusalem in 1939. His interests were in analysis and the history of mathematics.

Pal Turán (1910–1976) Turán was born in Budapest. He took his Ph.D. at the University of Budapest under Fejér. He taught at Budapest, working in number theory and graph theory.

Stanislaw Ulam (1909–1984) Born in Lvov, Poland, Ulam was educated at the Polytechnic Institute, Poland, and moved to the Institute for Advanced Study in 1936 for one year. He then held appointments at Harvard, Wisconsin, the Los Alamos Scientific Laboratory, the University of Colorado, and the University of Florida. His range of interests spanned various branches of pure mathematics and theoretical physics. He is probably best known for his work on the Manhattan Project.

James Victor Uspensky (1883–1947) Uspensky was born in Urga, Mongolia and educated at the University of St. Petersburg, receiving his Ph.D. there in 1910. He was a member of the Russian Academy of Sciences before joining the faculty at Stanford University in 1929. While in Russia, he taught one of the great twentieth-century number theorists, I. M. Vinogradov. Uspensky's fields were the theory of numbers, probability and analysis.

Oswald Veblen (1880–1960) The son of a physics professor at the University of Iowa, Veblen studied at Iowa, Harvard and Chicago. His principal appointment was at Princeton from 1905 till 1932 when he moved to the Institute for Advanced Study. He worked in topology, projective and differential geometry.

János (John) von Neumann (1903–1957) Born in Budapest, von Neumann received his undergraduate degree in chemical engineering from the Eidgenössische Technische Hochschule in Zürich and his doctorate at the University of Budapest. He received both undergraduate and graduate degrees the same year, 1926. He taught at Berlin, Hamburg and Princeton and was an early appointee to the Institute for Advanced Study. He worked in a wide variety of fields: analysis, applied mathematics, game theory and mathematical economics. Further, he was a pioneer in the development of computer science.

Karl Theodor Wilhelm Weierstraß (1815–1897) Weierstraß was born in Ostenfelde in Germany. He studied at the University of Bonn after which he taught in secondary schools. After receiving an honorary doctorate in 1854, he moved to Berlin and in 1864 he was given a professorship at the University of Berlin. He was one of the foremost analysts of the nineteenth century.

Hermann Klaus Hugo Weyl (1885–1955) Weyl was born in Elmshorn, Germany. He studied at the University of Göttingen, where he taught until 1913 when he went to Zürich. In 1930 he returned to Göttingen but in 1933 refused to remain in Germany, so he took a position at the Institute for Advanced Study in Princeton where he remained till he retired. His interests were very broad, spanning analysis, geometry, analytic number theory, topology, and quantum mechanics, among others.

Norbert Wiener (1894–1964) Wiener was born in Columbia, Missouri. He studied at Harvard (zoology), Cornell (philosophy) and Cambridge (mathematics with Russell), and when he returned to the United States he received an appointment to the Department of Mathematics at MIT, where he remained through the rest of his career. His research interests were in stochastic processes, quantum theory and cybernetics (a field which he named).

Wilhelm Wirtinger (1865–1945) Wirtinger was born in Ybbs, Austria. He received his Ph.D. from Vienna in 1887, after which he pursued studies in Berlin and Göttingen. He taught at the universities of Vienna and Innsbruck, working in a wide range of mathematical fields: function theory, number theory, geometry, invariants, as well as mathematical physics.

Photographs —
Acknowledgements

All illustrations are from the collection of the author unless otherwise noted.

Index